Springer Series in

OPTICAL SCIENCES 92

founded by H.K.V. Lotsch

Springer
Berlin
Heidelberg
New York
Hong Kong
London
Milan
Paris
Tokyo

Springer Series in
OPTICAL SCIENCES

The Springer Series in Optical Sciences, under the leadership of Editor-in-Chief *William T. Rhodes*, Georgia Institute of Technology, USA, provides an expanding selection of research monographs in all major areas of optics: lasers and quantum optics, ultrafast phenomena, optical spectroscopy techniques, optoelectronics, quantum information, information optics, applied laser technology, industrial applications, and other topics of contemporary interest.
With this broad coverage of topics, the series is of use to all research scientists and engineers who need up-to-date reference books.

The editors encourage prospective authors to correspond with them in advance of submitting a manuscript. Submission of manuscripts should be made to the Editor-in-Chief or one of the Editors. See also springeronline.com

Jukka Räty
Kai-Erik Peiponen
Toshimitsu Asakura

UV-Visible Reflection Spectroscopy of Liquids

With 131 Figures

Springer

Dr. Jukka Räty
University of Oulu
Measurement and Sensor Laboratory
Technology Park 127
87400 Kajaani
Finland

Professor Kai-Erik Peiponen
University of Joensuu
Department of Physics
80100 Joensuu
Finland

Professor Toshimitsu Asakura
Hokkai-Gakuen University
Department of Electronics
and Information Engineering
Minami 26 Nishi 11, Chuo-ku
Sapporo 064-0926, Hoikkaido
Japan

ISSN 0342-4111

ISBN 3-540-40582-8 Springer-Verlag Berlin Heidelberg New York

Cataloging-in-Publication Data:
Räty, Jukka A., 1958-, UV-visible reflection spectroscopy of liquids / Jukka A. Räty, Kai-Erik Peiponen, Toshimitsu Asakura. p.cm. – (Springer series in optical sciences, Issn 0342-4111;92) Includes bibliographical references and index. ISBN 3-540-40582-8 (acid-free paper) 1. Reflectance spectroscopy. 2. Liquids–Spectra. I. Peiponen, K.-E. (Kai-Eric) 1954- II. Asakura, Toshimitsu 1934- III. Title. IV. Springer series in optical sciences; v.92. QC454.R4R38 2004 530.4'2–dc22

Bibliographic information published by Die Deutsche Bibliothek Die Deutsche Bibliothek lists this publication in the Deutsche Nationalbibliografie; detailed bibliographic data is available in the Internet at <http://dnb.ddb.de>

Springer-Verlag is a part of Springer Science+Business Media

springeronline.com

Camera-ready by the author using a Springer TEX macropackage
Cover concept by eStudio Calamar Steinen using a background picture from The Optics Project. Courtesy of John T. Foley, Professor, Department of Physics and Astronomy, Mississippi State University, USA.
Cover production: *design & production* GmbH, Heidelberg

Printed on acid-free paper 56/3141/ts 5 4 3 2 1 0

To our families

Preface

Water is the most important element for a life. Scientists have been doing thorough investigations on the properties of water while engineers have exploited water in various industrial processes that aim at better optimization of the quality of industrial products. The knowledge of the physico-chemical properties of various liquids is extremely important in the field of life sciences. Recently, the study of liquids containing nanoparticles has been growing, since e.g. the modern development of new drugs by bioaffinity assays, and drug delivery is based more and more on the exploitation of nanoparticles in liquid phase. Optical metrology in context of drugs development is becoming an important tool in high-throuhgput screening of potential molecules. Unfortunately, process liquids in life sciences and in industrial environments are usually "ill-behaving" and it is usually difficult to monitor their condition. Fortunately, optical sensing based on reflectometry can be exploited in inspection of the quality of transparent and turbid liquids. This book is exceptional in the sense that it introduces the theory and practical reflectometry between same covers. We recommend this book for beginners and professionals who are doing research or working with a variety of liquids either in academic or engineering societies. The book is helpful for physicists, chemists and engineers since it covers topics related to basic physics and optics, optical spectra analysis, optical metrology and practical building of reflectometers.

The authors wish to express their cordial thanks to Professor Rauno Aulaskari for helping in the formulation of Appendix B and to Dr. Chun Ye for his critical comments. We will also remember with warmth Mrs. Riitta Honkanen and Mrs. Erja Koponen for helping to produce excellent drawings of this book. Likewise, Mr. Kyösti Karttunen, Mr. Ilpo Niskanen and Mr. Matti Räisänen receive our sincere thanks for their technical assistance. Finally, Dr. Räty is grateful to Mr. Juha Kalliokoski for his support during the course of writing this book.

Kajaani, Joensuu and Sapporo
November 2003

Jukka Räty
Kai-Erik Peiponen
Toshimitsu Asakura

Contents

Part I

Theory of Reflectometry

1 Demands on Measurement of Optical Constants of Liquids in Science and Industry

The optical properties of liquids have been investigated for a long time in science. Such investigations have provided much information related to the material properties of pure liquids. With the aid of optical spectroscopy the intrinsic optical properties of many liquids have been measured and calculated. Since the optical properties depend on the thermodynamic state of the liquid, refractive index and absorption of light has been investigated for liquids at different temperature and pressure. Such studies are usually related to basic studies in material sciences. Different optical measurement methods and theories for assessment of the optical constants (refractive index and extinction coefficient) have been well documented in literature. As an example of the richness of the methods and theory we mention the classical book written by Partington [1]. Nowadays, due to the progress in technology, such as machine vision, and in theory development, new measurement methods and theories have arisen, which may complement the older ones. One typical feature in modern optics is the demand on one hand that the optical constants are required with an increasing accuracy, and on the other hand smaller changes of the optical constants should be measured. Recently, the investigation of bio-optical liquids has become an important field of research in science and optical spectroscopy plays a big role in that field see e.g. [2]. As concern reflection spectroscopy in bio-optical research we mention for instance the utilization of surface plasmon resonance as a sensitive method for detection of dynamic biological interactions [3].

In industry where process liquids and liquid products are involved the refractive index is a basic quantity, which is used to the detection of the concentration of diluted solid material in various liquids. For instance, in water the contents of sugar, salt, proteins, acids etc. contribute to the refractive index of the solution. Traditionally the refractive index is given at 20°C for the sodium D-line wavelength equal to 589.3 nm. For the purpose of measurement of the refractive index of transparent liquids devices such as Abbe-refractometer have been developed and commercialized. Unfortunately, in process industry the liquids are usually strongly light absorbing and/or scattering. Then the conventional devices, such as Abbe-refractometer, become unreliable for the assessment of the refractive index of a turbid liquid. As an example of a turbid liquid we mention milk. Usually turbid liquids are diluted in order to exploit conventional refractometers for the estimation of the optical properties of the liquid, which means that the liquid sample has to be prepared for the measurement. However, if one wants to get real time data in industrial process line the preparation of the liquid is usually out of scope. Fortunately, it is possible to measure badly behaving liquids such as slurries using the principle of light reflection from an interface of two media. This opens the possibility to gain information both on dispersion and absorption of light of nontransparent liquids.

In addition to the industry there are also medical demands to measure e.g. various factors related to the blood etc. Thanks to the strong development of technology and the theory there are commercial devices, which operate in the light reflection mode and which have found also medical applications.

2 Liquids

2.1 Thermodynamics of Liquids

The liquid state can be defined as a medium, which has a fixed volume for a given amount of material, but not a fixed shape. Liquid state is usually considered to present a state between solids and gases. The common thing with solids is that liquids have relatively small molar volume and high density. Due to repulsive intermolar forces liquids are hard to compress. However, in liquids there is thermal energy, which allows the molecules to slide over each other. In other words there is no regular crystal lattice but the liquid can be optically considered to act as an amorphous solid.

Liquids share with gases the possibility to flow, and due to the thermal energy of molecules their motion is a random process such as in the case of gases. However, liquids take a fixed volume in a container, while gases expand to fill the whole container.

Any substance has a thermodynamic state independent on taking an equilibrium state or not. Furthermore, the laws of thermodynamics hold and they are not restricted only to equilibrium states. However, the equilibrium state is very important in the description of substances since it allows to present the state of the substance using well-defined physical quantities such as, temperature (T), pressure (p), volume (V) and mass (m). The importance of equilibrium states is that they can be unambiguously reproduced, thus being independent of time and place. The equation of state can be put on mathematical expression for a constant amount of substance as follows:

$$f(p, V, T) = 0 \, . \tag{2.1}$$

Obviously (2.1) presents a surface in the pVT-space. Due to the fact that a substance may appear either in solid, liquid or gaseous phase, one can expect that there is no simple explicit expression for (2.1).

Let us consider water, which is the most important substance for life and also most thoroughly studied substance in science. In Fig. 2.1 is shown the pVT-surface of pure water. From Fig. 2.1 we observe that the pVT-surface takes rather complex shape, although there appear regular sub areas. The complex shape is due to drastic changes in water, and in other substances also, when they experience phase transitions. As concerns the various thermodynamic (reversible) processes that can appear on the pVT-surface we refer

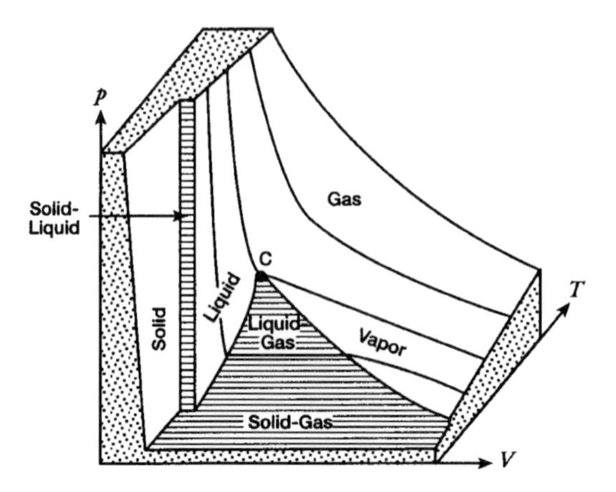

Fig. 2.1. pVT-surface of pure water. Point C is the critical point

only to the rich literature of thermodynamics. In the event that a single phase is present we are dealing with a homogenous system, whereas in the event that two or three different phases appear simultaneously in equilibrium the system is heterogeneous. Such systems are indicated on the pVT-surface in Fig. 2.1. Note that the point C in Fig. 2.1 is the critical point of the substance. The critical point presents the maximum pressure and temperature at which liquid and vapor can coexist in equilibrium. Above the critical point which in the case of water takes the values $p_c = 22\,\mathrm{MPa}$ and $T_c = 647.3\,\mathrm{K}$ there is no distinction between liquid and gas, therefore, the substance is called fluid. In the vicinity of the critical point the system has some dramatic changes in its optical properties which will be briefly described little bit later.

Next we express an equation that can be considered as the state equation of liquid phase. Suppose that we consider the volume of the liquid that depends on pressure and temperature. Then from (2.1) it follows that $V = V(T, p)$. Thus a differential change, dV, in the volume of the liquid (also solid) obeys the well-known relation

$$\frac{dV}{V} = \beta dT - \kappa dp \,, \tag{2.2}$$

where β is the isobaric volume expansivity defined by

$$\beta = \frac{1}{V}\left(\frac{\partial V}{\partial T}\right)_p \tag{2.3}$$

and κ is the isothermal compressibility defined by

$$\kappa = -\frac{1}{V}\left(\frac{\partial V}{\partial p}\right)_T \,. \tag{2.4}$$

Unfortunately, there is no theory that could provide mathematical expressions, of general validity, for β and κ. In general both of them depend on temperature and pressure and usually are positive numbers, which are obtained empirically. However, water is exception in the sense that κ takes negative values between 0 and 4°C.

2.2 Thermo-Optical Properties of Liquids

From the information of Fig. 2.1 we can qualitatively conclude that a relatively large increase of the pressure of the liquid (water in this case) has only a weak effect on the change of the volume. In other words the density of the water remains almost constant. However, when the temperature increases the liquid tends to expand, hence, the density of the liquid depends on the temperature. In optical physics we know that the variation of the density of a substance, such as liquid, induces also the variation of the complex refractive index of the substance, $N = n + \mathrm{i}k$, where n is the real refractive index and k is the extinction coefficient. The real refractive index and the extinction coefficient are usually termed as optical constants although they are not constants but depend on the thermodynamic state variables and the wavelength of the light. If we consider only the thermodynamic state variables of liquids the complex refractive index depends mainly on the temperature of the liquid and the contribution due to pressure change can usually be omitted. The remarkable thing with monitoring of the refractive index of a liquid (either stationary or flowing) is that information on density can be obtained in situ, which is not usually possible by other means. Furthermore, the accurate information of the complex refractive index of a liquid can be used to check the consistency of the state equations suggested for liquids.

Unfortunately, a general mathematical expression for the complex refractive index of liquids, as a function of temperature, meets the same problem as was discussed above in connection with (2.2)–(2.4). That is to say only empiric data is available and different parameters apply to different liquids. Naturally, the complex refractive index of the liquid depends on the wavelength of radiation incident upon it. Before going into details of spectral properties of liquids we present a figure analogous to that of water in Fig. 2.1, but now using optical properties of the different phases. This is shown in Fig. 2.2. We observe that water is transparent almost in every phase in the range of visible spectrum. However, the phase transitions have an effect on the refractive index of the water. Ice, which is transparent in the UV-VIS range, normally takes the hexagonal crystal structure at normal thermodynamic conditions. The hexagonal crystal structure means that optically uniaxial ice is birefringent. The birefringence is weak and ice has the lowest real refractive index from the known birefringent materials. The birefringence can be considered as a second order phase transition of water. The difference between liquid and solid water can be detected by noting that the dispersion and absorption

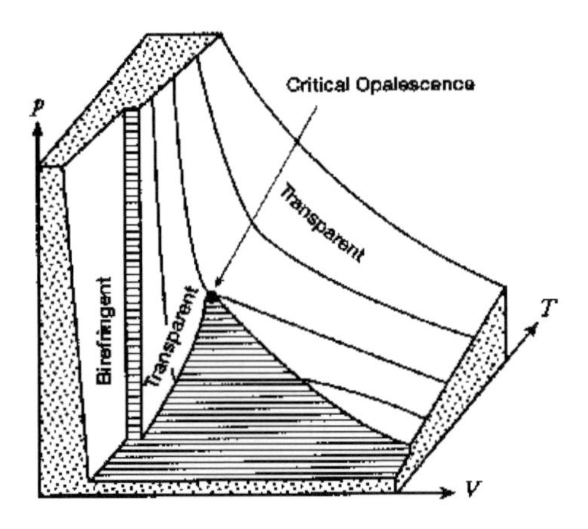

Fig. 2.2. Thermo-optical properties of pure water in pVT-space

of light in ice depends on the polarization degree of the incident light while in liquid water there is not such a dependence on the polarization state of the probe light.

In the case of liquid water the real refractive index is higher than in the case of ice. This is due to the higher density of the liquid phase. Water is optically isotropic so the polarization of the light is not a big issue as we already mentioned above. Spectroscopic properties of water and ice have been a subject of intensive studies [4–6]. In Fig. 2.3 are shown the real refractive index and extinction coefficients of water and ice at VIS-NIR spectral range. The ordinary and extraordinary optical constants of ice are practically speaking the same therefore two sets of optical constants are usually omitted especially in the event of bulk ice which is polycrystalline. The data of Fig. 2.3 can be obtained by measurement of reflectance from ice and water as a function of wavelength. The real refractive index and the extinction coefficient can be calculated with the aid of Kramers–Kronig analysis, which will be explained later in details in the chapter which is devoted to the spectra analysis.

As concerns the measurement of different parameters related to water quality in industry, and optical techniques involved we refer here only to [7].

The interesting feature with water and other liquids is the topology around the critical point. As we can observe from Figs. 2.1 and 2.2 there is a plateau on the isotherm passing through the critical point. This means that a very small temperature change from the critical value T_c has a large contribution to the critical volume V_c thus to the density of the system. Indeed, it is usually a problem to keep the system in equilibrium in the critical point. Actually, in the vicinity of the critical point there are large local fluctuations, on a scale of one micron, of the density of the fluid because the fluid near T_c tends to fluctuate between gaseous and liquid phases. Due to

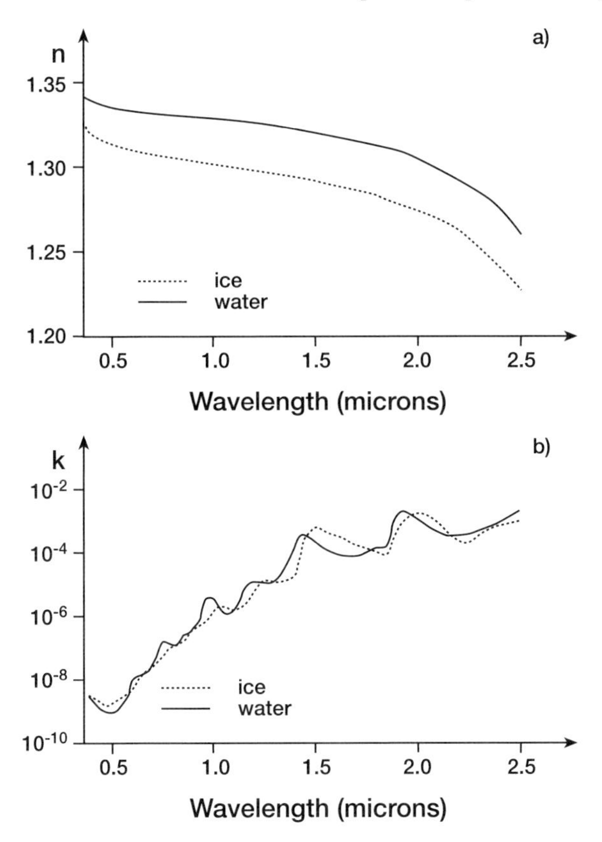

Fig. 2.3. Spectral properties of liquid water and ice. a) real refractive index, and b) extinction coefficient

the density fluctuation there appears also fluctuation of the refractive index of the fluid. This in turn causes strong scattering of light, which is called critical opalescence.

In the gaseous state water is transparent. The situation is different when we deal with a two-phase system (fog) of water droplets in air. Then the system usually strongly scatters light [8].

Above we have briefly described some basic thermodynamic and thermo-optic properties of liquids presenting pure substances. Unfortunately, most of the liquids that have importance in practical optical metrology either in engineering or in basic studies of science are rarely pure substances. Quite often we have to deal with multi-component mixtures such as liquid/liquid or liquid/solid systems. As an example of liquid/liquid system we mention the mixture of water and alcohol. The case of a liquid/solid system may deal with a solid material such as salt or sugar, which are diluted e.g. into the water, or Latex particles, which form a binary system with the water. Nevertheless, also then one can seek for equation of state, and optical constants as a function of state variables. However, their properties usually become more complicated due to the multi-component nature of the mixture and due to possible chemical reactions. In the study of liquids by reflectometry we face the problems of optically very dense liquids i.e. strongly absorbing but not scattering or more generally turbid liquids which can simultaneously absorb and scatter the incident light. As an example of a turbid liquid we mention milk which is opaque at the UV-VIS spectral range. Despite of the complexity of the liquids involved, which usually means also richness in their optical properties, one has to keep in mind especially that their optical constants depend on the thermodynamic state, although the state variables are usually taken as parameters. This fact has consequences for instance in industrial measurement environments where e.g. the temperature of a process liquid is quite often a subject of fluctuations. Naturally, there are other disturbances present in industry, which may affect the measurement such as turbulent flow, contamination of the probe window, dust on the surfaces of optical elements of the sensor etc. Similar disturbances can occur also in laboratory conditions when we wish to measure optical properties of liquid samples that have biological or medical origin.

3 Theory of Optical Constants

In the UV–VIS spectral range the interaction of electric field with a liquid phase is important and the interaction of magnetic field with liquid can usually be neglected in most cases of practical optical metrology. The electric charges of molecules in liquids experience the coulombian interaction. In addition to mutual interaction of the molecules, external electric field, such as light can disturb the charge distribution of the electrons. Liquids belong to materials, which are called dielectrics. Dielectrics are non-conducting but polarizable media, which can be either permanently polar or nonpolar. Molecules, which have a center of symmetry i.e. they are arranged in a symmetric manner are nonpolar, whereas asymmetric molecules are polar. However, light can disturb the electron charge cloud of a polar or nonpolar medium and, hence, to induce a polarization. Next we consider in details the interaction of light with the electronic system of a liquid.

3.1 Interaction of Light with Liquid

Since we are working at the UV–VIS spectral range we can neglect orientational polarization, which arises from rotation of permanent dipole moments of asymmetric molecules. The interaction of UV–VIS light with liquids is related to electronic polarization, which depends on the displacement of electrons with respect to the nucleus of an atom. In other words the external light field distorts the orbits of the negatively charged electrons that move around the positively charged nucleus. This induced electronic polarization of charges occurs in all dielectrics. Let us consider a single electron that is bound to a nucleus as shown in Fig. 3.1. Note that at this stage we are ignoring, which however will be taken into account a little bit later, the interaction of the system with the molecules surrounding the system of Fig. 3.1. The electron of the dielectric is object to damped harmonic motion striven by the external electric light field $E = E_0 \exp(-\mathrm{i}\omega t)$, where E_0 is the amplitude of the external electric field, ω is the angular frequency of radiation, t denotes time and i is the imaginary unit. The equation of motion of the electron is as follows:

$$m\frac{d^2x}{dt^2} + m\Gamma\frac{dx}{dt} + bx = -eE_0e^{-\mathrm{i}\omega t}\,, \tag{3.1}$$

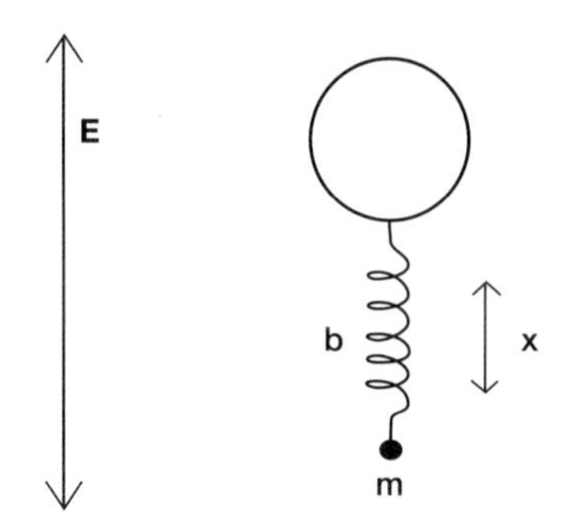

Fig. 3.1. Electric-field-driven oscillator model for insulators. The electric field propagates from left to right

where m is the mass of the electron, $-e$ is the electron charge, Γ is the damping parameter and b is the spring coefficient. The damping parameter and the spring coefficient have also quantum mechanical interpretations, and for more details about their quantum nature we refer to the book of Wooten [9]. We wish to emphasize that both Γ and b depend on the temperature of the liquid. Naturally, in general case, there are oscillators that have motion that is not restricted only in the direction of x-axis, and in the case of optically anisotropic liquids (liquid crystals) the spring coefficients depend on the direction of the incident light. However, since here we wish to give a qualitative picture about the light interaction with liquid, then the description bellow is sufficient. The solution of the second-order differential equation (3.1) can be found by a trial function

$$x(t) = x_0 e^{-\mathrm{i}\omega t} , \tag{3.2}$$

where x_o is the maximum amplitude of the oscillation. Substitution of the trial function into (3.1) gives the solution, which is

$$x(t) = \frac{eE_0}{m} \frac{e^{-\mathrm{i}\omega t}}{\omega_0^2 - \omega^2 - \mathrm{i}\Gamma\omega} , \tag{3.3}$$

where $\omega_0 = (b/m)^{1/2}$ is the natural angular frequency, i.e. the frequency of the oscillator without damping. Naturally, we have to accept only the solution (3.3), which presents real numbers. However, the complex form (3.3) is important since it allows us to take simultaneously into account both dispersion and absorption of light.

The dipole moment (p) of the system of Fig. 3.1 can now be expressed as follows:

$$p(t) = -ex(t) = \frac{(e^2 E_0/m)e^{-\mathrm{i}\omega t}}{\omega_0^2 - \omega^2 - \mathrm{i}\Gamma\omega} . \tag{3.4}$$

Then the macroscopic polarization P can be written with the aid of number density ρ of the electrons:

$$P = \rho p . \tag{3.5}$$

The microscopic polarizability a is defined by the relation

$$a = \frac{p}{E} \ . \tag{3.6}$$

From the theory of electromagnetism we know that the macroscopic polarization can be expressed with the aid of linear susceptibility $\chi^{(1)}$ and the external electric field as follows:

$$P(t) = \epsilon_0 \chi^{(1)} E(t) \ , \tag{3.7}$$

where ϵ_0 is the permittivity of vacuum. The linear susceptibility carries information about the interaction of light with medium, which can be a dielectric, metal or semiconductor. If we get information by measurement on the linear susceptibility of a liquid then we have the possibility to characterize its microscopic properties. The permittivity of the dielectric medium is be given as follows:

$$\epsilon = \epsilon_0 (1 + \chi^{(1)}) \ . \tag{3.8}$$

With the aid of (3.4), (3.5), (3.7) and (3.8) we can equate the real and imaginary parts of the relative Lorentzian permittivity $\epsilon_r = \epsilon / \epsilon_0$ and obtain

$$\begin{aligned} \mathrm{Re}\{\epsilon_r(\omega)\} &= 1 + \frac{\rho e^2}{m\epsilon_0} \frac{\omega_0^2 - \omega^2}{\left(\omega_0^2 - \omega^2\right)^2 + \Gamma^2 \omega^2} \\ \mathrm{Im}\{\epsilon_r(\omega)\} &= \frac{\rho e^2}{m\epsilon_0} \frac{\Gamma \omega}{\left(\omega_0^2 - \omega^2\right)^2 + \Gamma^2 \omega^2} \ . \end{aligned} \tag{3.9}$$

In Fig. 3.2 are shown typical curves calculated from (3.9). The real part of (3.9) describes dispersion and the imaginary part absorption of light. From historical reasons the angular frequency range $[\omega_0 - \Gamma/2, \omega_0 + \Gamma/2]$ is called the range of anomalous dispersion.

Next we define the complex refractive index N with the aid of complex permittivity and permeability as follows:

$$N^2 = \mu_r \epsilon_r = \left(\frac{\mu}{\mu_0}\right) \left(\frac{\epsilon}{\epsilon_0}\right) \ . \tag{3.10}$$

Since we are dealing with nonmagnetic medium we can set $\mu_r = 1$. Thus we can write

$$\begin{aligned} \mathrm{Re}\{\epsilon_r\} &= n^2 + k^2 \\ \mathrm{Im}\{\epsilon_r\} &= 2nk \ . \end{aligned} \tag{3.11}$$

In the UV–VIS range we can get directly information only on the complex refractive index but not on the complex permittivity. The complex permittivity can be calculated with the aid of (3.11). In theoretical considerations when we know the complex permittivity of a dielectric and wish to calculate

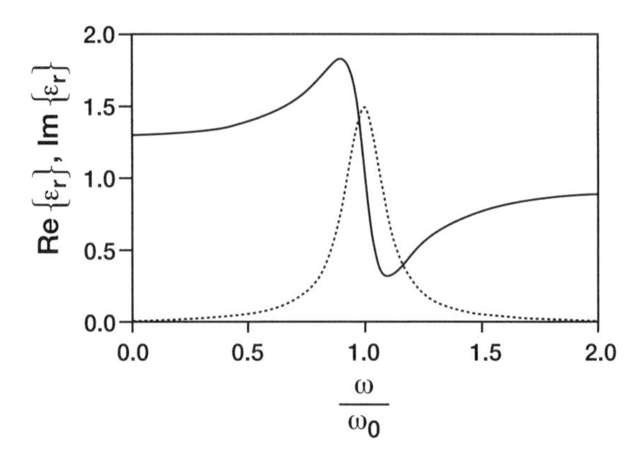

Fig. 3.2. An example of real (solid line) and imaginary (dotted line) parts of relative Lorentzian permittivity. The angular frequency scale is dimensionless since it is normalized to the resonance frequency

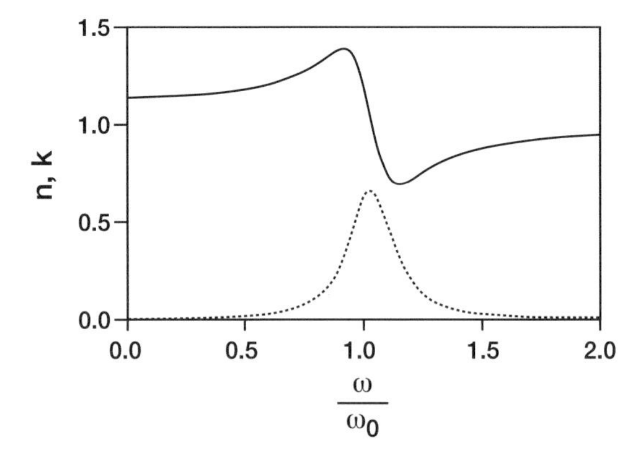

Fig. 3.3. Real refractive index (solid line) and extinction coefficient (dotted line) calculated using the data of Fig. 3.2

the refractive index and the extinction coefficient, the following formulas are valid

$$
\begin{aligned}
n &= \left(\frac{1}{2}\left\{\left[(\mathrm{Re}\{\epsilon_r\})^2 + (\mathrm{Im}\{\epsilon_r\})^2\right]^{1/2} + \mathrm{Re}\{\epsilon_r\}\right\}\right)^{1/2} \\
k &= \left(\frac{1}{2}\left\{\left[(\mathrm{Re}\{\epsilon_r\})^2 + (\mathrm{Im}\{\epsilon_r\})^2\right]^{1/2} - \mathrm{Re}\{\epsilon_r\}\right\}\right)^{1/2} .
\end{aligned}
\tag{3.12}
$$

In Fig. 3.3 are plotted the refractive index and the extinction coefficient obtained using the data of Fig. 3.2.

Unfortunately, the real and imaginary parts of the relative complex permittivity of liquids do not satisfy the simple relations of (3.9). One reason is that we neglected in our derivation the dipole–dipole interaction. In a more appropriate model for electronic polarization we have to take into account the other dipoles which affect to the electric field strength experienced by the dipole of Fig. 3.1. Lorentz considered the local field of the dipole on a basis of a model where the dipole is at the center of a sphere in a dielectric [9]. The radius of the sphere is taken large enough as compared with the dimensions of the molecules. Some of the dipoles are inside and some are outside the sphere. By calculating (details of calculations can be found e.g. in [9] and [10]) the electric field components of the dipoles one can get so-called Clausius–Mossotti equation as follows:

$$\frac{\epsilon_r - 1}{\epsilon_r + 2} = \frac{\rho a}{3\epsilon_0} . \tag{3.13}$$

Clausius–Mossotti equation is the basis of the approximation of the real refractive index of many liquids including water [5]. If we replace the complex relative permittivity of the liquid by that of the complex refractive index we obtain another practical formula which is known as the Lorentz–Lorentz equation:

$$\frac{N^2 - 1}{N^2 + 2} = \frac{\rho a}{3\epsilon_0} . \tag{3.14}$$

In (3.14) it is possible to take into account thermal motion of polar liquids with the aid of the Boltzmann's distribution law from thermodynamics. Then a term on right side of (3.14) equal to $\rho p^2/(9\rho_0 KT)$ has to be added, where K is the Boltzmann's constant. The resulting equation is known as Debye's equation. Since the frequency of light at UV–VIS range is relatively high the Lorentz–Lorentz equation can be used even for polar liquids. Quite often a practical form of (3.14) is that of molar refraction, which is only a weak function of temperature and density, but depends on the wavelength of the incident light. This formula is

$$\frac{N^2 - 1}{N^2 + 2} \frac{M}{D} = \frac{N_A a}{3\epsilon_0} , \tag{3.15}$$

where M is the molecular mass, D is the density of the liquid, and N_A is the Avogadro's number. A Sellemeier type dispersion formula for weakly absorbing liquids can be exploited in the description of the wavelength dependence of the molar refraction [11]

$$\frac{n^2 - 1}{n^2 + 2} \frac{M}{D} = C_0 + \sum_j \frac{C_j}{\lambda^2 - \lambda_j^2} , \tag{3.16}$$

where λ_j:s are the resonance wavelengths of the liquid and C_j:s are constants related to the optical material properties of the liquid in question. An empiric formula for water, based on (3.16), which takes into account spectral features, the temperature and the density of the water, was given by Sciebener et al. [5]:

$$\frac{n^2-1}{n^2+2}\frac{1}{D^*} = 0.243905091 + 9.53518094 \cdot 10^{-4} D^* - 3.64358110 \cdot 10^{-3} T^{*2} + 2.65666426 \cdot 10^{-4} \lambda^{*2} T^* + 1.59189325 \cdot 10^{-3} / \lambda^{*2} + \frac{2.45733798 \cdot 10^{-3}}{\lambda^{*2} - 0.229202^2} + \frac{0.897478251}{\lambda^{*2} - 5.432937^2} - 1.63066183 \cdot 10^{-2} D^{*2} , \qquad (3.17)$$

where $D^* = D/1000\,\mathrm{kg/m^3}$, $\lambda^* = \lambda/0.589\,\mathrm{nm}$ and $T^* = T/273.15\,\mathrm{K}$. The equation (3.17) can be used to assess the refractive index of water from $-10°\mathrm{C}$ to $+500°\mathrm{C}$ in temperature, 0 to $1045\,\mathrm{kg/m^3}$ in density and 0.2 to $2.5\,\mu\mathrm{m}$ in wavelength.

We remark that the resonance frequency of the Lorentz–Lorentz system is different from the resonance frequency of (3.9). In many cases the Lorentz–Lorentz equation is not sufficient if we want to resolve accurately the optical constants of transparent liquids. The simple model fails usually also for turbid liquids. However, there are rather general mathematical procedures, which make it possible to resolve the optical constants from measured data as will be shown later when we consider spectra analysis in more details. In such cases we can computationally reconstruct the wavelength-dependent complex refractive index of a liquid or other substance.

If we deal with a mixture of transparent liquids, such as water and alcohol, the real refractive index can be obtained from [12]

$$n = n_{SL} + f(n_{SL} - n_{sl}) , \qquad (3.18)$$

where n_{SL} is the refractive index of the solvent, n_{sl} is the refractive index of the solute and f is the volume fraction of the solute.

3.2 Light Interaction with Optically Nonlinear Kerr Liquid

In optical spectroscopy and when using a tunable high intensity dye laser as a light source nonlinear optical processes of liquids and other media can be studied. In the event that the isotropic liquid is optically nonlinear (actually all isotropic media possess third-order nonlinearity) the polarization of the electrons depends on the strength of the electric field of the incident light. The relation between the polarization and the electric field can be given using the concept of nonlinear susceptibility $\chi^{(n)}$, where n is the order of the nonlinear process. The relation can be given with a series expansion

$$P(E) = P(0) + \epsilon_0(\chi^{(1)} E + \chi^{(2)} E^2 + \chi^{(3)} E^3 + \dots) , \qquad (3.19)$$

where the first term $P(0)$ describes the permanent polarization of polar medium, the second term is the linear response of medium, whereas the higher terms in the series expansion are so-called hyperpolarizabilities. Most often the liquids are optically isotropic then there are no intrinsic second-order nonlinear processes present. In the event that anistropic molecules are present in the liquid a polarized electric field of intense light can cause reorientation of the molecules, hence light-induced birefringence. Before dealing with the nonlinear contribution of the refractive index we consider briefly some basic physics related to nonlinear optical phenomena. For this purpose let's start with the equation of motion of an electron. Because the interaction of light with the dipole of Fig. 3.1 is now dependent on the strength of the electric field we have to modify the equation of motion of the electron. This is accomplished by introducing the equation of motion of an anharmonic oscillator as follows:

$$\frac{d^2x}{dt^2} + \Gamma\frac{dx}{dt} + \omega_0^2 x + \sum_{j=2}^{\infty} b_j x^j = -\frac{e}{m}E_0 e^{-\mathrm{i}\omega t} , \tag{3.20}$$

where the sum $\sum b_j x^j$ includes the anharmonic terms. Unfortunately, the differential equation (3.20) is nonlinear, which usually means complications in seeking for solutions. However, in this case the solution can be found with the aid of a trial function, which is given by the series expansion

$$x = \sum_{l=1}^{\infty} x_l E^l . \tag{3.21}$$

After some iterative algebra [13] one obtains the solutions for x_l. Bellow we give the three first solutions, by considering only two first terms in the sum of the anharmonic terms in (3.20), which are

$$\begin{aligned}
x_1 &= -\frac{e}{m}\frac{E_0 e^{-\mathrm{i}\omega t}}{\omega_0^2 - \omega^2 - \mathrm{i}\Gamma\omega} \\
x_2 &= \frac{e^2 b_2^2}{m^2}\frac{E_0^2 e^{-2\mathrm{i}\omega t}}{(\omega_0^2 - 4\omega^2 - 2\mathrm{i}\Gamma\omega)(\omega_0^2 - \omega^2 - \mathrm{i}\Gamma\omega)^2} \\
x_3 &= -\frac{e^3}{m^3}\left(\frac{2b_2^2}{\omega_0^2 - 4\omega^2 - 2\mathrm{i}\Gamma\omega} + b_3\right)\frac{E_0^3 e^{-3\mathrm{i}\omega t}}{(\omega_0^2 - \omega^2 - \mathrm{i}\Gamma\omega)^3(\omega_0^2 - 9\omega^2 - 3\mathrm{i}\Gamma\omega)} .
\end{aligned} \tag{3.22}$$

The solution x_1 presents the case of linear optics, whereas x_2 and x_3 present generation of harmonic waves due to nonlinear processes. Now the polarization, in turn, is obtained from

$$P = -\rho e \sum_{l=1}^{\infty} x_l E^l . \tag{3.23}$$

If we substitute x_l into (3.23) and equate (3.19) and (3.23) we obtain solutions for the second- and third-order nonlinear susceptibilities in the case of harmonic wave generation. There are various nonlinear processes that involve other frequencies than those of harmonic waves, in general

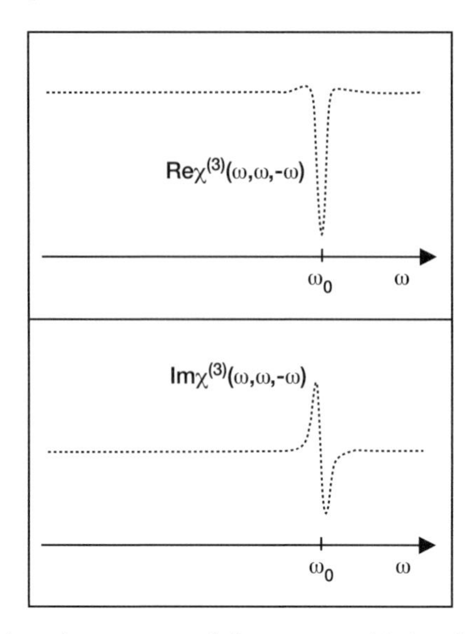

Fig. 3.4. Real and imaginary parts of degenerate third-order nonlinear susceptibility as a function of angular frequency. The real part has minimum at resonance frequency

$\chi^{(n)} = \chi^{(n)}(\omega_1, \omega_2, \ldots, \omega_n)$. This can be justified by introducing a driving electric field $\sum E_l e^{-i\omega_l t}$ on right-hand side of (3.20). Here, we are only interested in the complex refractive index related to the degenerate nonlinear susceptibility, which can be expressed as a function of the angular frequency of incident light that is to say the important quantity is $\chi^{(3)} = \chi^{(3)}(\omega, \omega, -\omega)$. Then we are dealing with a self-action process. Now the liquid is considered as a Kerr medium (e.g. nitrobenzene) and its complex third-order nonlinear susceptibility is the following in our description [13]:

$$\chi^{(3)}(\omega, \omega, -\omega) = \rho \frac{e^4}{m^3 \epsilon_0} \left(b_3 + \frac{2b_2^2}{3} \left[D(2\omega) + 2D(0) \right] \right) (D(\omega))^3 D(-\omega) , \tag{3.24}$$

where $D(\omega) = (\omega_0^2 - \omega^2 - i\Gamma\omega)^{-1}$. The real and imaginary parts of (3.24) are plotted in Fig. 3.4. The real part somehow resembles the "absorption curve" and the imaginary part the "dispersion curve", in linear optics.

The total complex refractive index of optically nonlinear liquid or other substance can be defined as:

$$N(\omega, \omega, -\omega; I) = N_L(\omega) + N_{NL}(\omega, \omega, -\omega; I) , \tag{3.25}$$

where N_L is the linear complex refractive index, N_{NL} is the contribution of the complex nonlinear refractive index and I is the intensity of the light. The real parts of the complex linear and nonlinear refractive indices constitute the total refractive index. The connection between the real part of the nonlinear refractive index and the third-order nonlinear susceptibility of isotropic medium is as follows [14]:

$$\mathrm{Re}\left\{N_{NL}(\omega,\omega,-\omega)\right\} = \frac{3}{4\epsilon_0 c\left[\mathrm{Re}\left\{N_L(\omega)\right\}\right]^2}\mathrm{Re}\left\{\chi^{(3)}(\omega,\omega,-\omega)\right\}I\,, \quad (3.26)$$

where c is the light velocity in vacuum. Quite often the total real refractive index is presented simply in the form

$$n = n_L + n_{NL}I\,. \quad (3.27)$$

The Lorentz–Lorentz formula (3.14) can be generalized to hold also in the frame of (3.27).

The imaginary part of the nonlinear refractive index is related to two-photon absorption and it holds that [14]

$$\mathrm{Im}\left\{N_{NL}(\omega,\omega,-\omega)\right\} = \frac{3\omega}{2\epsilon_0 c^2\left[\mathrm{Re}\left\{N_L(\omega)\right\}\right]^2}\mathrm{Im}\left\{\chi^{(3)}(\omega,\omega,-\omega)\right\}I\,. \quad (3.28)$$

A strong pump beam can also affect on the complex refractive index related to a weak probe beam, which both have different angular frequencies. Then we usually are involved with Raman induced Kerr effect (RIKES) see e.g. [15].

3.3 Beer–Lambert Law

In the case of liquids that are optically linear the light absorption obeys the familiar intensity law:

$$\frac{dI}{dx} = -\alpha I\,. \quad (3.29)$$

After integration of the linear first-order differential equation (3.29) we get the well-known Beer–Lambert law

$$I = I_0 e^{-\alpha(\lambda)d}\,, \quad (3.30)$$

where I_0 is the incident light intensity, α is the absorption coefficient, λ is the wavelength of the light and d is the thickness of the liquid sample. The light attenuation is illustrated in Fig. 3.5. According to (3.30) light intensity is exponentially attenuated in the medium while it travels through the medium. The law of (3.30) is the basis of spectrophotometers that are used in measurement of transmission of light through absorbing media. Sometimes it is possible to make corroborative measurements using a transmission spectrophotometer in order to check the data obtained in reflection spectroscopy.

The connection between the absorption and extinction coefficient is:

$$k(\lambda) = \frac{\lambda\alpha(\lambda)}{4\pi}\,. \quad (3.31)$$

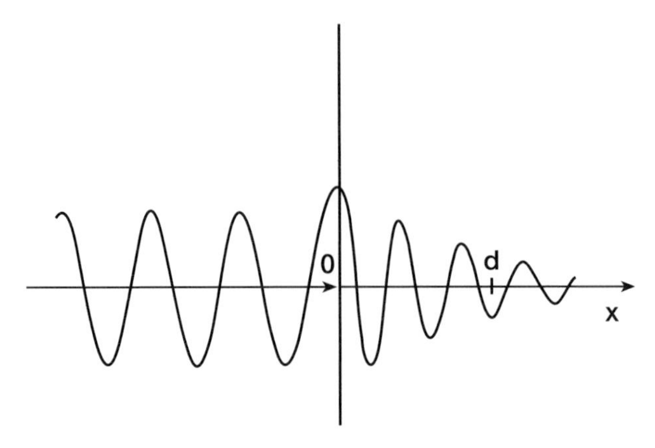

Fig. 3.5. Electric field attenuation in medium. The medium is located on the right-hand side of the vertical axis

For a Kerr liquid, which is illuminated with a single light beam, the probability of two-photon absorption is more probable than other multi-photon absorption processes. Then the nonlinear differential equation taking into account two-photon absorption is given by [14]

$$\frac{dI}{dx} = -\alpha I - \gamma I^2 , \tag{3.32}$$

where α is the linear absorption coefficient and γ is the absorption coefficient of two-photon absorption. Two-photon absorption coefficients for several liquids at UV region were measured by Dragomir et al. [16]. The solution of (3.32) is presented in details in Appendix A here we give the result which yields

$$I = \frac{\frac{\alpha}{\gamma}}{\left(1 + \frac{\alpha}{\gamma I_0}\right) e^{\alpha x} - 1} . \tag{3.33}$$

The total absorption coefficient is usually put in a simple formula, analogous to (3.27),

$$\alpha = \alpha_L + \alpha_{NL} I . \tag{3.34}$$

We will consider the nonlinear optical properties of liquids, but containing nanoparticles, again in section 3.5.

3.4 Effective Medium Theory

In modern technology the exploitation of nanoparticles have drawn much interest. As concerns the present state-of-art we refer here to the WTEC Panel Report on Nanostructure Science and Technology [17]. Nanoparticles

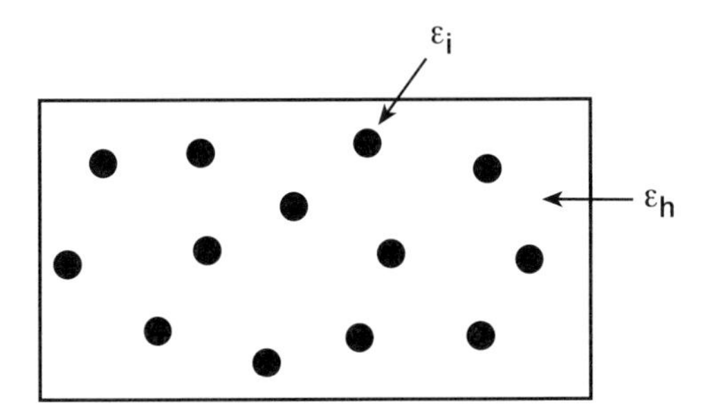

Fig. 3.6. A model for Maxwell–Garnett nanocomposite. The black circles denote inclusions embedded in the host medium

in liquid environments have importance, since they can be exploited e.g. in controlled drug delivery [18] and in drug development with the aid of bioaffinity assays using two-photon absorption [19, 20]. Optical research of biomolecules in aqueous solutions can be expected to play important role in medicine.

We consider the effective medium theory of liquids containing optically linear and later also optically nonlinear nanoparticles. This means usually that we can neglect light scattering, especially if the volume concentration of the nanoparticles is low and they don't form aggregates i.e. there is no agglomeration. The basis of the treatment has its origin on research that was started at the beginning of 20th century. Maxwell Garnett [21, 22] and Bruggeman [23] were the pioneers in formulation of the effective medium theory. We take some time to describe a Maxwell Garnett nanosphere-liquid two-phase system, and thereafter briefly describe the corresponding Bruggeman system. The Maxwell Garnett model resembles to great extent the model of Lorentz–Lorentz in Sect. 3.1. In Fig. 3.6 is presented a schematic diagram of a two-phase Maxwell Garnett nanoparticle system. The solid nanospheres can either be insulators, metals or semiconductors. The host material is liquid and nanospheres are inclusions. We assume that there is no interaction between the inclusions. This means e.g. that the fill fraction (f) of the nanospheres is relatively low $f \ll 1$. The upper limit of the fill fractions has been considered to be as high as $f = 0.5$ in studies related to nanocomposites [24]. We deal with the effective permittivity (ϵ_{eff}) of the system and hint here again to the classical theory of electromagnetism [25] and especially a nice pedagogic paper of Aspnes [26], where the derivation of the effective medium expression of permittivity for Maxwell Garnett two-phase nanocomposite can be found. It holds that

$$\frac{\epsilon_{\text{eff}} - \epsilon_h}{\epsilon_{\text{eff}} + 2\epsilon_h} = f \frac{\epsilon_i - \epsilon_h}{\epsilon_i + 2\epsilon_h} , \tag{3.35}$$

where ϵ_h is the complex permittivity of the host and ϵ_i is the complex permittivity of the inclusion. After some tedious but straight forward algebra one can resolve the real and imaginary parts of the complex effective permittivity as follows [27]:

$$\epsilon'_{\mathrm{eff}} = \frac{\Psi\left[(1+2f)(\epsilon'_i\epsilon'_h - \epsilon''_i\epsilon''_h) + 2(1-f)(\epsilon'^2_h - \epsilon''^2_h)\right]}{\Psi^2+\Phi^2} - \frac{\Phi\left[(1+2f)(\epsilon'_i\epsilon''_h + \epsilon''_i\epsilon'_h) + 4(1-f)\epsilon'_h\epsilon''_h\right]}{\Psi^2+\Phi^2}$$

$$\epsilon''_{\mathrm{eff}} = \frac{\Phi\epsilon'_{\mathrm{eff}} + (1+2f)(\epsilon'_i\epsilon''_h + \epsilon''_i\epsilon'_h) + 4(1-f)\epsilon'_h\epsilon''_h}{\Psi}\,, \tag{3.36}$$

where

$$\Psi = (1-f)\epsilon'_i + (2+f)\epsilon'_h$$
$$\Phi = -(1-f)\epsilon''_i - (2-f)\epsilon''_h \tag{3.37}$$

and the prime indicates the real part and two primes the imaginary part of the effective permittivity. In Fig. 3.7 are shown the real and imaginary parts of effective relative permittivity calculated from (3.36) and (3.37) for different fill fractions and under the assumption that $\epsilon_h = 1.77$, and the complex relative permittivity of the inclusions is the one obeying (3.9). In optical spectroscopy we usually get information on the effective complex refractive index (N_{eff}) of the present system. Then it is possible to get the effective permittivity of the liquid with the aid of (3.11). Furthermore, if ϵ_h and f are known one can obtain information on ϵ_i from (3.38)

$$\epsilon_i = \frac{\epsilon_h[2(\epsilon_{\mathrm{eff}} - \epsilon_h) + f(\epsilon_{\mathrm{eff}} + 2\epsilon_h)]}{f(\epsilon_{\mathrm{eff}} + 2\epsilon_h) - (\epsilon_{\mathrm{eff}} - \epsilon_h)}\,. \tag{3.38}$$

Since we usually wish to find out the complex refractive index of the inclusions i.e. $N_i = n_i + k_i$, we have to solve n_i and k_i using (3.38) and (3.12). It is also possible to monitor f if the permitttivities of the host and inclusions both are known.

The shortcoming of the Maxwell Garnett effective medium theory is that the fill fraction should be low. In addition there is no information about the size parameter of the nanosphere. Maxwell Garnett model can be extented [28] to include the concept of size parameter e.g. on the basis of the Mie scattering theory [8]. The Maxwell Garnett effective medium theory has been applied also in the case of intrepretation of ellipsometric data related to microscopic surface roughness in spectroscopy [29], and in connection with anisotropic composite [30]. Maxwell Garnett effective medium theory makes difference between the host and the inclusions, whereas Bruggeman gave a theory where neither phase were given a preference but dealing with inclusions that are embedded in the effective medium itself. Then the role of fill fraction is not so critical as in the case of Maxwell Garnett system. For a randomly intermixed Bruggeman nanosphere-liquid system it holds

$$f_a\frac{\epsilon_a - \epsilon_{\mathrm{eff}}}{\epsilon_{\mathrm{eff}} + g(\epsilon_a - \epsilon_{\mathrm{eff}})} + f_b\frac{\epsilon_b - \epsilon_{\mathrm{eff}}}{\epsilon_{\mathrm{eff}} + g(\epsilon_b - \epsilon_{\mathrm{eff}})} = 0\,, \tag{3.39}$$

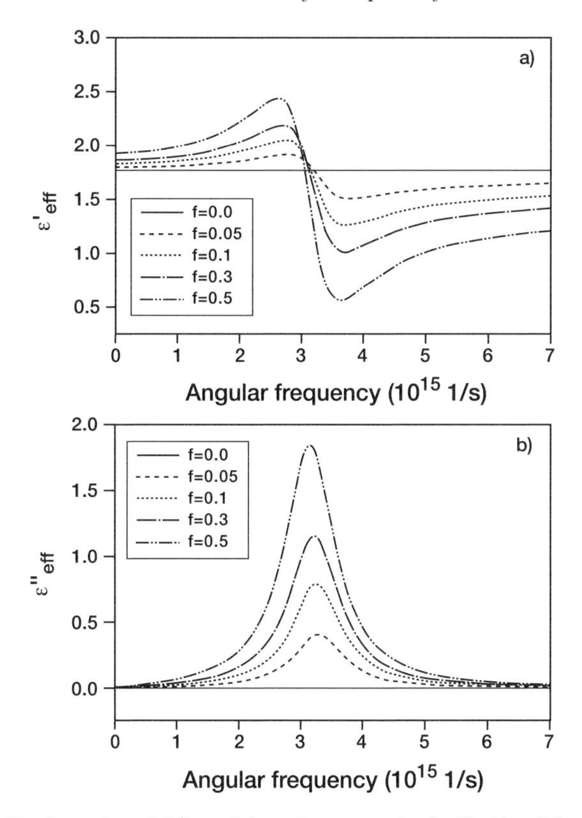

Fig. 3.7. a) Real part and b) and imaginary part of effective Maxwell–Garnett effective permittivity as a function of angular frequency and the fill fraction

where a and b denote the two components having different complex permittivities and fill fractions, and g is a geometric factor which depends on the shape of the "inclusions". For "spherical inclusions" $g = 1/3$ and for "two-dimensional circular inclusions" $g = 1/2$. Also in this model the effective relative permittivity and the effective complex refractive index of the liquid can be coupled using (3.11) or (3.12).

3.5 Effective Medium Theory of Optically Nonlinear Liquids

If optically linear nanospheres are in optically nonlinear liquid matrix, and the refractive index of the nanospheres is lower than the refractive index of the liquid, then a strong enhancement of a nonlinear process from the liquid can be obtained. The reason for strong enhancement is the collective enhancement of the local electric fields in the vicinity of nanospheres and in general of nanostructures. The enhancement of nonlinear optical signal from nanocomposites has been given experimental evidence [31, 32]. The study of nonlinear optical properties of nanospheres and coated nanospheres, has been an object of thorough investigations [33–36]. Sipe and Boyd [37] presented the formulation of the mathematical expressions for cases where the Maxwell Garnett inclusions are permitted to be optically nonlinear while the host (liquid in our case) acts in linear manner and vice versa. They also gave the theory for a case where both the inclusions and the host medium are optically nonlinear [37]. Dispersion theory of nonlinear susceptibilities of Maxwell Garnett medium was presented by Peiponen et al [13, 24, 38]. Next we give the experssion of Sipe and Boyd [37] for the degenerate third-order nonlinear effective susceptibility in the case where both the inclusions and the host are optically nonlinear

$$\chi_{\mathrm{eff}}^{(3)}(\omega,\omega,-\omega) = f\left(\frac{\epsilon_{\mathrm{eff}}(\omega)+2\epsilon_h(\omega)}{\epsilon_i(\omega)+2\epsilon_h(\omega)}\right)^2 \left|\frac{\epsilon_{\mathrm{eff}}(\omega)+2\epsilon_h(\omega)}{\epsilon_i(\omega)+2\epsilon_h(\omega)}\right|^2 \chi_i^{(3)}(\omega,\omega,-\omega)$$
$$+\left(\frac{\epsilon_{\mathrm{eff}}(\omega)+2\epsilon_h(\omega)}{3\epsilon_h(\omega)}\right)^2 \left|\frac{\epsilon_{\mathrm{eff}}(\omega)+2\epsilon_h(\omega)}{3\epsilon_h(\omega)}\right|^2 [(1-f)+xf]\chi_h^{(3)}(\omega,\omega,-\omega)\,, \tag{3.40}$$

where $\chi_i^{(3)}$ is the nonlinear susceptibility of the inclusions, $\chi_h^{(3)}$ the nonlinear susceptibility of the host, f is the fill fraction of inclusions, and

$$x = \frac{8}{5}\eta^2|\eta|^2 + \frac{6}{5}\eta|\eta|^2 + \frac{2}{5}\eta^3 + \frac{18}{5}(|\eta|^2+\eta^2) \tag{3.41}$$

$$\eta = \frac{\epsilon_i(\omega)-\epsilon_h(\omega)}{\epsilon_i(\omega)+2\epsilon_h(\omega)}\,. \tag{3.42}$$

Figure 3.8 illustrates, for the shake of simplicity, the real and imaginary parts of (3.40) under the assumption that $\chi_h^{(3)} = 0$ and

$$\chi_{\mathrm{eff}}^{(3)}(\omega,\omega,-\omega) = \frac{C}{|\omega_0^2-\omega^2-\mathrm{i}\Gamma\omega|^2(\omega_0^2-\omega^2-\mathrm{i}\Gamma\omega)^2}\,, \tag{3.43}$$

where C is a constant. In this case there is no enhancement of the effective nonlinear susceptibility, however, the spectroscopic properties of the system depend on the light intensity. From Fig. 3.8a we observe that there are two positions for zero dispersion. Obviously the zero dispersion on left-hand side can be tuned by the fill fraction so that it will shift towards lower energy as f increases. In optical spectroscopy we may try to estimate the value of fill fraction, under the assumption of present model, by observing the position of the zero dispersion. From Fig. 3.8a we observe also that the strongest change in the real part of the effective nonlinear susceptibility is in the vicinity of the resonance angular frequency. The imaginary part, Fig. 3.8b, of the effective nonlinear susceptibility is becoming strongest for angular frequencies higher than the resonance angular frequency. The position of the minimum

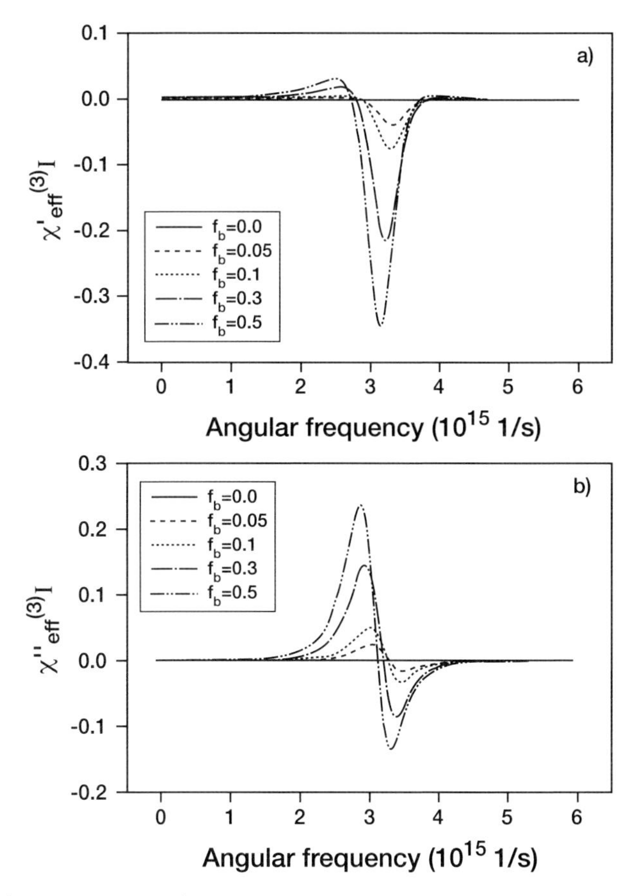

Fig. 3.8. a) Real part and b) imaginary part of the Maxwell–Garnett susceptibility as a function of angular frequency and fill fraction

or maximum of the imaginary part depends on the fill fraction in Fig. 3.8b. In Fig. 3.9 are shown the real and imaginary parts of the effective linear and total permittivities. Evidently the nonlinear nanospheres in liquid can have an effect on the spectral properties of the real and imaginary parts of the total permittivity of Fig. 3.9. Same holds also for the total complex refractive index of the nonlinear Maxwell Garnett liquid. In principle, the contribution of the nonlinear refractive index can provide, in addition to the linear optical properties, information about the fill fraction and optical properties of the nonlinear nanospheres. This possibility is important if there appear simultaneously two types of nanospheres, which have as an average the same size parameter and the same linear refractive index, but the other type of the sphere is optically linear while the other is optically nonlinear. In such a case we have a three-phase system and the theoretical treatment is little bit more complex than for a two-phase system.

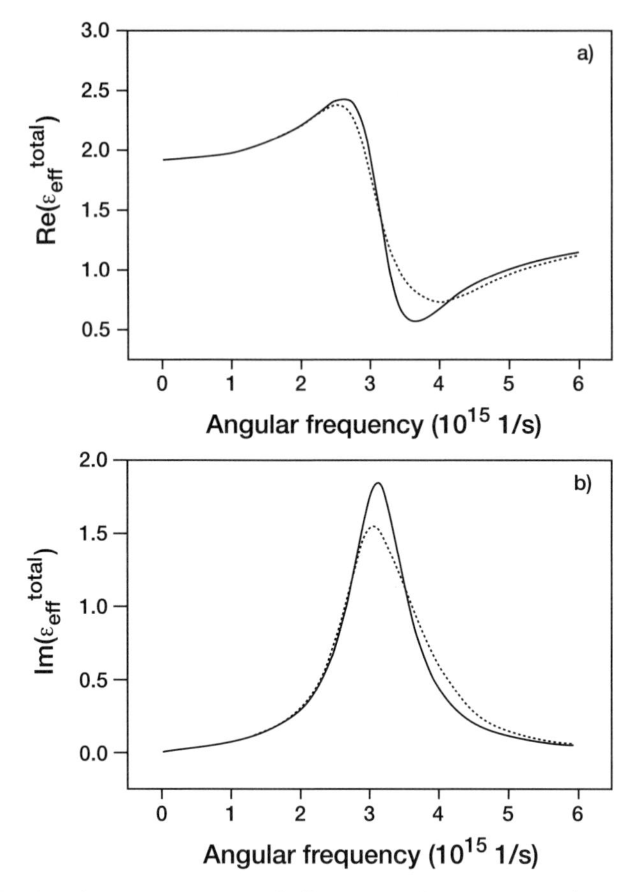

Fig. 3.9. Real and imaginary parts of effective total permittivity (solid curves) and linear permittivity (dotted curves). The curves we calculated using data of Figs. 3.8 and 3.9 for $f = 0.5$

In the case of nonlinear Bruggeman system the effective nonlinear third-order susceptibility is given by the approximation [24] and [33]

$$\chi_{\text{eff}}^{(3)} = \sum_j \frac{1}{f_j} \left| \frac{\partial \epsilon_{\text{eff}}}{\partial \epsilon_j} \right| \left(\frac{\partial \epsilon_{\text{eff}}}{\partial \epsilon_j} \right) \chi_j^{(3)} , \qquad (3.44)$$

where the index j indicates the j:th constituent. The Bruggeman model (3.44) has been applied e.g. in the studies where a liquid fills the pores of porous glass [39] and in description of discontinuous thin liquid films [40].

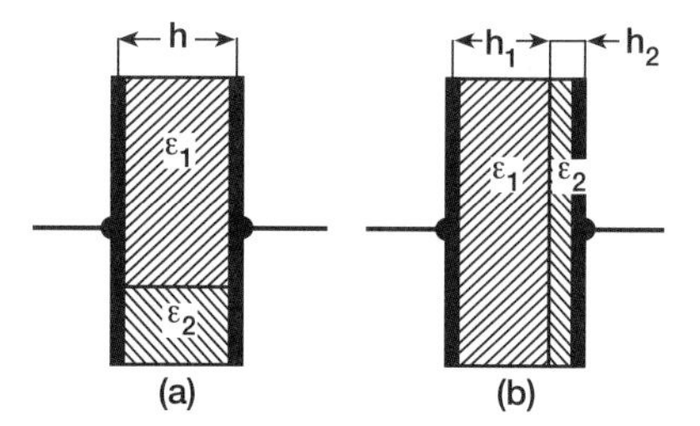

Fig. 3.10. Capacitors used in modelling the effective permittivity, a) two parallel dielectrics and b) two series connected dielectrics

3.6 Wiener Inequalities

Above we were dealing with rather simple Maxwell Garnett and Bruggeman effective medium theory. In many cases, especially for industrial process liquids the particles in the liquid, which can be of nano or microsize, take usually other shape than that of a sphere. Indeed, the particles may resemble e.g. a needle or they are irregular. Then the above models for effective medium usually fail. Nevertheless, there is usually a desire to measure the fill fraction or the spectral properties of the irregular shape particles. This is rather tough problem, especially if the sample is dense slurry. Fortunately, there is rather simple method to find upper and lower bounds for the complex effective permittivity or refractive index of the matrix. This can be accomplished using Wiener inequalities [41]. Let us consider the capacitors of Fig. 3.10. Inside the capacitors there are two parallel and series connected dielectrics. In the case of Fig. 3.10a the total capacitance C_{tot} of the capacitor is as follows:

$$C_{\text{tot}} = C_1 + C_2 = \frac{\epsilon_1 S_1}{d} + \frac{\epsilon_2 S_2}{d} = \frac{\epsilon_{\text{eff}}(S_1 + S_2)}{d} , \tag{3.45}$$

where d is thickness of the dielectric and S_1 and S_2 denote the areas of the dielectrics on the electrodes. According to the last equality of (3.45) we consider as if the capacitor would be filled with a single dielectric which has an effective permittivity equal to ϵ_{eff}. From (3.45) one can solve that

$$\epsilon_{\text{eff}} = \epsilon_1 \frac{S_1}{S_1 + S_2} + \epsilon_2 \frac{S_2}{S_1 + S_2} . \tag{3.46}$$

Since the area ratios in (3.46) are equal to the fill fractions f_1 and f_2, we can write

$$\epsilon_{\text{eff}} = f_1 \epsilon_1 + f_2 \epsilon_2 . \tag{3.47}$$

In the case of the capacitor of Fig. 3.10b for series connection the total capacitance is

$$\frac{1}{C_{\text{tot}}} = \frac{1}{C_1} + \frac{1}{C_2} , \tag{3.48}$$

where now

$$C_1 = \frac{\epsilon_1 S}{d_1}$$
$$C_2 = \frac{\epsilon_2 S}{d_2} . \tag{3.49}$$

This means that the total capacitance obeys the formula

$$C_{\mathrm{tot}} = \frac{\epsilon_{\mathrm{eff}} S}{d_1 + d_2} . \tag{3.50}$$

The areas of both dielectrics is the same on the electrode, which means that the fill fractions in this case can be given by

$$f_1 = \frac{d_1}{d_1 + d_2}$$
$$f_2 = \frac{d_2}{d_1 + d_2} . \tag{3.51}$$

It follows from (3.48)–(3.51) that

$$\frac{1}{\epsilon_{\mathrm{eff}}} = \frac{f_1}{\epsilon_2} + \frac{f_2}{\epsilon_2} . \tag{3.52}$$

It is evident from (3.46) and (3.52) that they can be generalized to hold for j various dielectrics. Wiener concluded that the true effective real permittivity of any statistic mixture should lie between the following inequalities

$$\frac{1}{\sum_{j=1}^{J} \frac{f_j}{\epsilon_j}} \leq \epsilon_{\mathrm{eff}} \leq \sum_{j=1}^{J} f_j \epsilon_j . \tag{3.53}$$

Wiener bounds can be applied also to the effective complex refractive index and they can be applied in the case of nonlinear materials as well by introducing the appropriate permittivities, which take into account the intensity dependencies.

As an example let us deal with concentric nonabsorbing spherical shell in nonabsorbing slurry [42] shown in Fig. 3.11. We wish to estimate the unknown refractive index (as a function of the wavelength) of the shell material n_{shell}, while we have measured the effective refractive index n_{eff} of the slurry using a reflectometer. Furthermore, we assume that we know the refractive index of the liquid n_{liq} (e.g. in the case of water equation (3.17)) and the effective refractive index of the particle that is to say the refractive index of the shell plus liquid inside the shell. In Sect. 3.7 we give expressions that can be used in estimation of the refractive index of the particle n_p. It is assumed that same liquid is inside and outside the shell. Furthermore, we assume that the outer and inner diametres of the shell does not vary to a great extent. In

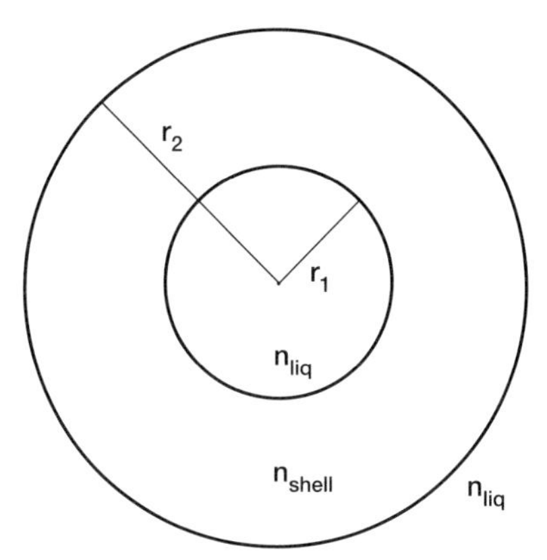

Fig. 3.11. A pigment particle constituting of concentric spherical shells. Liquid is assumed to fill the inner cavity of the spherical pigment particle. Same liquid is asumed to occupy the exterior of the particle

practical cases, such as e. g. plastic pigments which have a concentric spherical shell, and which are used as raw materials in paint and paper industry, the diameters vary but usually their average values can be used in calculations. According to (3.53), we get the lower and upper bounds (*lb* and *ub*)

$$n_{\text{eff,lb}} = \sqrt{\frac{1}{\frac{f_{\text{liq}}}{n_{\text{liq}}^2} + \frac{f_p}{n_p^2}}} \leq n_{\text{eff}} \leq \sqrt{f_{\text{liq}} n_{\text{liq}}^2 + f_p n_p^2} = n_{\text{eff,ub}} \,, \tag{3.54}$$

where f_{liq} and f_p are the known volume fill fractions of the liquid and the particles, respectively. The bounds of (3.54) can be used for testing the validity of the measured n_{eff}. If the measured effective refractive index is between the bounds of (3.54) then we can suppose that the measurements are consistent with the Wiener theory. Next if we know the fill fraction of the shell and the core, we can estimate the refractive index of the shell simply using the bounds

$$n_{\text{shell,lb}} = \sqrt{\frac{n_p^2 - f_{\text{core}} n_{\text{liq}}^2}{f_{\text{shell}}}} \leq n_{\text{shell}} \leq \sqrt{\frac{f_{\text{shell}}}{\frac{1}{n_p^2} - \frac{f_{\text{core}}}{n_{\text{liq}}^2}}} = n_{\text{shell,ub}} \,. \tag{3.55}$$

The utility of such bounds will be demonstrated later, in Part II of this book when we deal with reflection spectra obtained from plastic pigment slurry. An important tool for such a study is transmission electron microscope, which provides information about the diameters of the concentric spherical shell.

3.7 Mie Scattering from Turbid Liquid

Industrial liquids contain many types and irregular shape solid particles, whose size can vary in a large scale and in addition they can birefringent. This makes the estimation of the refractive index of the turbid liquid quite problematic. Usually we can get rather easily an optical signal from such media, but the interpretation of the measurement result is problematic because of the lack of appropriate theory. In the event that the particles in the liquid are spherical, optically isotropic, and their size is around the wavelength of the light Mie scattering theory can be applied [8], [43] and [44] for the assessment of the refractive index and extinction coefficient of dilute dispersions. The underlying assumption is that the light absorption of the dispersion is negligible. Thus the extinction is due to the scattering of light from the liquid that contains spherical particles. Maltsev et al [45] have developed a method for real-time determination of size and refractive index of individual microspheres in turbid but weakly absorbing liquid by using so-called flying light scattering indicatrix.

According to Van de Hulst [43] the complex refractive index of a dispersion has the following expression

$$\begin{aligned} N &= n' + \mathrm{i}n'' \\ &= n_{\mathrm{liq}} + \frac{n_{\mathrm{liq}}\lambda_0^3\rho}{4\pi^2}\mathrm{Im}\{S(0°)\} + \mathrm{i}\frac{n_{\mathrm{liq}}\lambda_0^3\rho}{4\pi^2}\mathrm{Re}\{S(0°)\}\,, \end{aligned} \quad (3.56)$$

where n_{liq} is the refractive index of the pure liquid and $S(0°)$ is the complex forward scattering amplitude of a single particle. This scattering amplitude is obtained from Mie theory, which states that

$$S(0°) = \frac{1}{2}\sum_{j=1}^{\infty}(2j+1)(a_j + b_j)\,. \quad (3.57)$$

The Mie coefficients are obtained from the following formulas

$$\begin{aligned} a_j &= \frac{m\Psi_j(mx)\Psi_j'(x) - \Psi_j(x)\Psi_j'(mx)}{m\Psi_j(mx)\xi_j'(x) - \xi_j(x)\Psi_j'(mx)} \\ b_j &= \frac{\Psi_j(mx)\Psi_j'(x) - m\Psi_j(x)\Psi_j'(mx)}{\Psi_j(mx)\xi_j'(x) - m\xi_j(x)\Psi_j'(mx)}\,, \end{aligned} \quad (3.58)$$

where m is the ratio of the refractive index of the spherical particle to that of the liquid, $x = \epsilon_{\mathrm{liq}}^{1/2}\omega D/2c$, D is the diameter of the particle, and Ψ_j and ξ_j are Riccati–Bessel functions. It has been suggested that the model of (3.56) can be applied also to high concentrations of the particles [46–48] and linear dependence of the refractive index with respect to the volume fractions up to close-packing has been observed [46–48]. The intresting feature is that then n' of a turbid liquid depends only on the single-particle properties.

Mohammadi [49] presented a simple formula for the approximation of the effective refractive index of a non-absorbing suspension,

$$n' = n_{\text{liq}} + f(n_p - n_{\text{liq}})\frac{\sin X}{X} \tag{3.59}$$

and

$$X = \frac{4.5(n_p - n_{\text{liq}})D}{\lambda_0}\,, \tag{3.60}$$

where n_p is the refractive index of the particle. The equations (3.59) and (3.60) are valid [49] up to the first zero $X = \pi$ of the oscillating function $\sin X/X$.

In the case of turbid liquid that absorbs and scatters light the situation is somewhat more complicated than above. The light extinction in turbid liquid, and in forward scattering direction, is governed by the Beer–Lambert law

$$I(z) = I_0 e^{-(\alpha_{\text{abs}}+\alpha_{\text{sca}})z}\,, \tag{3.61}$$

where α_{abs} is related to absorption, α_{sca} to scattering and z denotes distance. In the case of transmission measurement, information of light attenuation related to absorption and scattering can be obtained, but in reflection measurement from turbid liquids Meeten and North [47] conjectured that when absorption and scattering coexist the reflectance signal carries information dominantly due to absorption and refraction of light. The functional dependence of n'' on volume fraction of the scattering particles in turbid liquid has been observed to be nonlinear [46–48], which makes its utility in industrial measurement environment usually less practical.

4 Theory of Reflectance

The accurate measurement of the refractive index of transparent liquids is possible using conventional devices such as Abbe-refractometer or an interferometer. The problems of estimation of the real refractive index arise when the liquid is optically very thick i.e. absorbing but non-scattering, or the liquid is turbid which usually means relatively strong scattering of light and sometimes simultaneous absorption of light. Fortunately, reflection spectroscopy can be applied to such difficult tasks that require the estimation of the complex refractive index of turbid liquids. In addition reflection spectroscopy can provide means for adsorption studies [50] such as for example contamination of probe window. There are sources that describe thoroughly the exploitation of reflection spectroscopy in material research [51, 52] and industrial measurement applications [53]. However, there has been recent progress both in technology for reflection spectroscopy and spectra analysis, which give new light to the physical phenomena behind the measured data related to reflection measurements. In this chapter we will describe the theory of the oblique incidence light reflection from absorbing or turbid liquids.

4.1 Fresnel's Formulas for Oblique Light Incidence

Fresnel's formulas are of great importance for conducting reflectance experiments at the interface of two homogenous media. Suppose a monochromatic plane wave E_i is arriving at the interface of two media shown in Fig. 4.1.

We assume that the lower medium in Fig. 4.1 is liquid, which has a complex refractive index N_2 while the upper medium is the prism material with real refractive index n_1 (absorption of prism material can usually be neglected). The prism is needed for the detection of the light reflection from the liquid and the dispersion of the prism is usually known. In Fig. 4.1 are shown the s- and p-polarized electric field components, the s-component is the one, which is perpendicular to the plane of incidence and the p-component parallel to the plane of incidence. We make use here the Verdet convention [54]. The laws of electromagnetism together with boundary conditions can be used to determine the reflected E_r and transmitted E_t electromagnetic waves. Furthermore, the amplitude reflection coefficients r_s and r_p can be established in the following manner [11]:

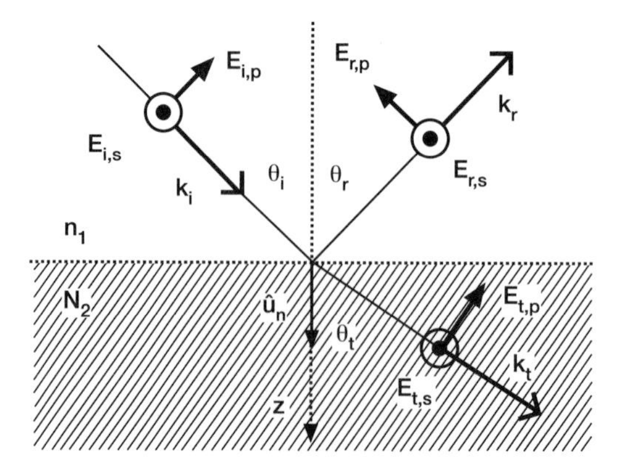

Fig. 4.1. An electromagnetic wave at the interface of two semi-infinite media. E_i, E_r and E_t are incident, reflected and transmittted electric field vectors. Suscripts s and p represent field components: s-normal and p-paralleel to the plane of incidence. Wave vectors are indicated as k_i, k_r and k_t

$$r_s = \frac{\cos\theta - (N_{21}^2 - \sin^2\theta)^{1/2}}{\cos\theta + (N_{21}^2 - \sin^2\theta)^{1/2}} \tag{4.1}$$

and

$$r_p = \frac{N_{21}^2\cos\theta - (N_{21}^2 - \sin^2\theta)^{1/2}}{N_{21}^2\cos\theta + (N_{21}^2 - \sin^2\theta)^{1/2}} , \tag{4.2}$$

where θ is the angle of incidence and $N_{21} = N_2/n_1$. From now on we denote N_{21} by N and the refractive index of the prism, if explicit needed, by $n_1 = n_{\text{prism}}$. Equations (4.1) and (4.2) are Fresnel's formulas for the electric field amplitude applying to all linear, isotropic and homegenous nonmagnetic media. They hold both for external, $Re\{N\} > 1$, and to internal reflection, $Re\{N\} < 1$. The internal reflection is important in the field of the present book, which is devoted to the optical spectroscopy of liquids. We remark that it is possible, not only in external, but also in internal reflection spectroscopy that the modulus $|N/n_{\text{prism}}| > 1$. It is obvious from (4.1) and (4.2) that r_s and r_p are complex numbers, which can be given by

$$\begin{aligned} r_s &= |r_s|e^{i\varphi_s} \\ r_p &= |r_p|e^{i\varphi_p} . \end{aligned} \tag{4.3}$$

The phase angles φ_s and φ_p describe the phase shift of the electric field during the reflection from the interface of the two media. Since the amplitude reflection coefficients depend on the relative complex refractive index by (4.1) and (4.2), and the relative complex refractive index is a function of the wavelength of the incident light, the phase angles are functions of wavelength and usually depart from the values 0 or π denoted in classical optics. The phase angles can be calculated from reflectance using a phase retrieval procedure. Such a method will be described in details in Chap. 6.

Reflectometers can not detect the phase of the electric field, but the intensity, which means that we get information about the reflectance from the liquid. This in turn means that the Fresnel's formulas for the reflectance obey the following familiar expressions

$$R_s = \left| \frac{\cos\theta - (N^2 - \sin^2\theta)^{1/2}}{\cos\theta + (N^2 - \sin^2\theta)^{1/2}} \right|^2 \tag{4.4}$$

$$R_p = \left| \frac{N^2\cos\theta - (N^2 - \sin^2\theta)^{1/2}}{N^2\cos\theta + (N^2 - \sin^2\theta)^{1/2}} \right|^2 . \tag{4.5}$$

Equations (4.4) and (4.5) form the basis of various kind of analysis of reflectance as pointed out by Humphreys–Owen [55]. In this book we concentrate on the methods where the angle of incidence, s-and p-polarization, and especially scanning of wavelength are exploited in determination of the complex refractive index of an isotropic liquid. Nevertheless, optical constants can be resolved also, using appropriate Fresnel's formulas, for birefringent liquids such as nematic liquid crystals [56, 57].

4.2 Assam's Formulas

Azzam has carried out comprehensive theoretical studies on the reflection of polarized light [58–62]. He provided a practical relation [59] between the amplitude reflectivities of s- and p-polarized light. Futhermore, Azzam gave practical relations for the phase angle of reflectivity of s- and p-polarized light valid for oblique light incidence [59]. We start by defining a parameter q by the relation

$$q = \sqrt{N^2 - \sin^2\theta} . \tag{4.6}$$

Then according to (4.1) and (4.2) it holds that

$$r_s = \frac{\cos\theta - q}{\cos\theta + q} \tag{4.7}$$

and

$$r_p = \frac{(q^2 + \sin^2\theta)\cos\theta - q}{(q^2 + \sin^2\theta)\cos\theta + q} . \tag{4.8}$$

From (4.7) it follows that

$$q = \left(\frac{1 - r_s}{1 + r_s} \right) \cos\theta . \tag{4.9}$$

We substitute the result of (4.9) into (4.8) and get the result of Azzam

$$r_p = r_s \left(\frac{r_s - \cos 2\theta}{1 - r_s \cos 2\theta} \right) . \tag{4.10}$$

Equation (4.10) holds for any two isotropic media that define the planar interface and throughout the whole electromagnetic spectrum.

If we take the squared modulus on both sides of (4.10) we find out that

$$R_p = R_s \frac{R_s + \cos^2 2\theta - 2\sqrt{R_s}\cos 2\theta \cos\phi_s}{1 + R_s \cos^2 2\theta - 2\sqrt{R_s}\cos 2\theta \cos\phi_s} . \quad (4.11)$$

From the algebraic equation (4.11) we can solve

$$\cos\varphi_s = \frac{R_s^2 - R_p + R_s(1 - R_p)\cos^2\theta}{2\sqrt{R_s}(R_s - R_p)\cos 2\theta} . \quad (4.12)$$

Using (4.10) and (4.12) it is possible to find also the relation [59]

$$\tan(\varphi_p - \varphi_s) = \frac{\sqrt{R_s}\sin^2 2\theta \sin\varphi_s}{\sqrt{R_s}\cos\varphi_s(1 + \cos^2 2\theta) - R_s \cos 2\theta} . \quad (4.13)$$

The simplest way to obtain the complex refractive index of a liquid is based on the reflectometric measurement of both R_s and R_p for a fixed angle of incidence, fixed wavelength, and on calculation based on (4.12) and the following relation

$$N_{\text{liquid}} = n_{\text{prism}} \sqrt{\sin^2\theta + \cos^2\theta \left(\frac{1 - r_s}{1 + r_s}\right)^2} . \quad (4.14)$$

The validity of the Azzam's method can be estimated with the aid of the error of the reflectance and the angle of incidence. As an example we consider a simulation presented by Räty and Peiponen [63]. The simulation is based on the use of the following inequality, which is related to the modulus of partial differentials of the complex refractive index

$$\triangle N \le \sum \left| \frac{\partial N}{\partial x_j} \right| \triangle x_j , \quad (4.15)$$

where the variables x_j are the reflectances R_s and R_p and the angle of incidence. In Fig. 4.2 an example of error analysis, using (4.15), for simulated media ($n_1 = 1.500$, $n_2 = 1.350$ and $k_2 = 0.002$) is shown. The measurement errors of $\triangle\theta = 0.01°$ and $\triangle R_s = \triangle R_p = 0.5 \cdot 10^{-3}$ were introduced. According to Fig. 4.2 it is obvious that the choice of angle of incidence has a significant effect on the results relating to optical constants.

4.3 Brewster- and Pseudo-Brewster Angle

In Figs. 4.3a and b are shown the internal reflectances R_s and R_p as a function of the angle of incidence for a nonabsorbing and absorbing medium for s- and p-polarized light. From Figs. 4.3a and b we can observe that the reflectance of p-polarized light has a minimum at a certain angle of incidence.

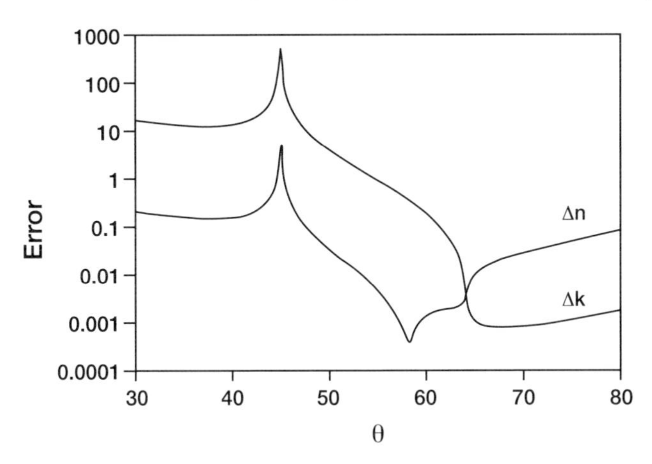

Fig. 4.2. Error analysis of the Azzam method for a simulated medium. The analysis employs parameters for the two media: $n_1 = 1.5000$, $n_2 = 1.350$ and $k_2 = 0.002$

Moreover, the minimum is zero for nonabsorbing medium (Fig. 4.3a) and nonzero for absorbing medium (Fig. 4.3b). The latter property is not so evident from Fig. 4.3b, but it can be distinguished from Fig. 4.3c, which illustrates the case of strongly absorbing medium. Let us consider the simpler case of nonabsorbing medium. It is obvious from (4.5) that the zero is obtained if

$$N^2 \cos\theta = \sqrt{N^2 - \sin^2\theta}\,. \tag{4.16}$$

From (4.16) we can equate

$$N^4 \cos^2\theta = N^2 - 1 + \cos^2\theta\,, \tag{4.17}$$

which, in turn, yields that

$$\cos^2\theta = \frac{1}{N^2 + 1}\,. \tag{4.18}$$

Now because

$$\sin^2\theta = 1 - \cos^2\theta = \frac{N^2}{N^2 + 1} \tag{4.19}$$

we can solve from (4.18) and (4.19) that

$$\tan\theta_B = N\,, \tag{4.20}$$

where θ_B is the Brewster angle. Since N is a complex number in (4.20) we must require that

$$\begin{aligned} \tan\theta_B = \mathrm{Re}\{N\} &= n \\ \mathrm{Im}\{N\} = k &= 0\,. \end{aligned} \tag{4.21}$$

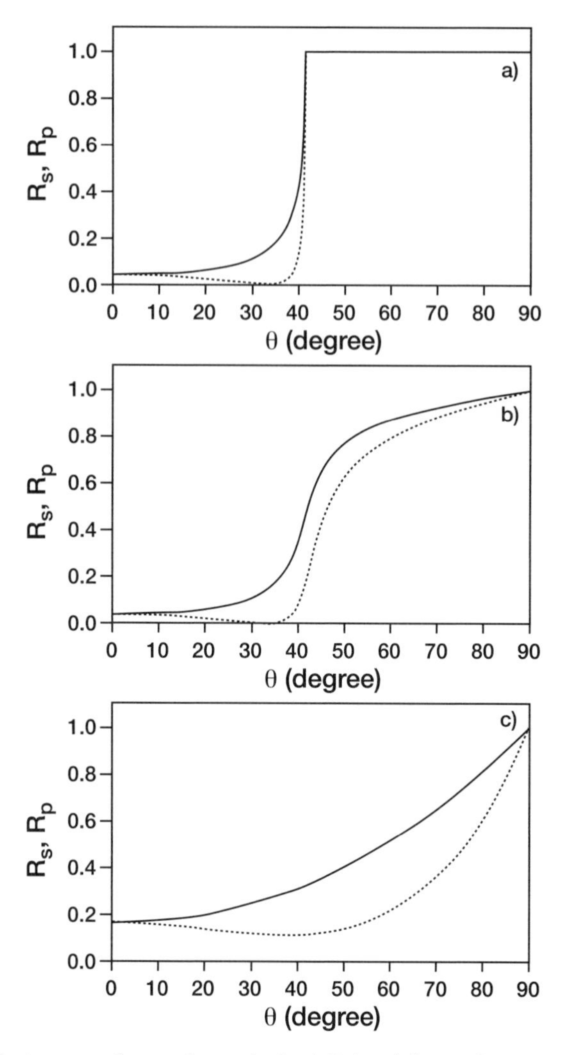

Fig. 4.3. Reflectances of s- and p-polarized light. (**a**) no absorption, (**b**) low absorption and (**c**) high absorption coefficient of a liquid

In the case of an absorbing liquid (i.e. $k > 0$), the minimum reflectance is nonzero, the corresponding equation (4.21) would deal with an apparent real refractive index of absorbing liquid. The corresponding angle of minimum reflectance is the pseudo-Brewster angle, which is a complicated function of n and k: The pseudo-Brewster angle is larger than the normal Brewster angle [64]. The location of the pseudo-Brewster angle can be obtained either by finding numerically the minimum of the curve $dR_p/d\theta$, or roots of the Humphreys–Owen's [55] equation

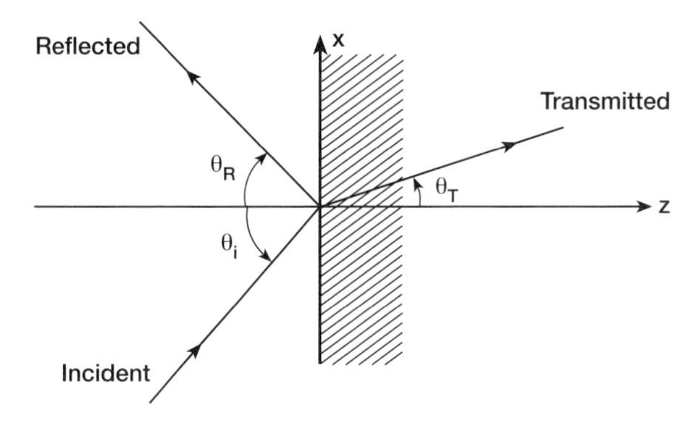

Fig. 4.4. Light reflection and transmission at an interface of two medium

$$P^4 \cos^4 \theta - P^2(1 + 2\cos^2 \theta) + 2Q \sin^2 \theta = 0 \, , \tag{4.22}$$

where

$$\begin{aligned} P &= (\mathrm{Re}\{N\})^2 + (\mathrm{Im}\{N\})^2 \\ Q &= (\mathrm{Re}\{N\})^2 - (\mathrm{Im}\{N\})^2 \, . \end{aligned} \tag{4.23}$$

Naturally the spectroscopic properties of the liquid rule the position of the Brewster and pseudo-Brewster angles. The pseudo-Brewster angle method has been shown to be relatively accurate for the estimation of the true real refractive index of absorbing and heterogenous samples [65].

4.4 Total Reflection from Nonabsorbing Liquids

The phenomenon of total reflection is justified in classical optics with the aid of the Snell law. If we look at the reflection curves for nonabosorbing medium in Fig. 4.3a we notice that at certain angle of incidence the reflectance abruptly jumps to a constant value, which is equal to unity. This angle is the critical angle of total reflection. By inspection we observe from (4.4) and (4.5) that the reflectance of s- or p-polarized light is equal to unity if the angle of incidence is equal to grazing incidence, i.e. $\theta = \pi/2$ or

$$\sqrt{N^2 - \sin^2 \theta} = 0 \, . \tag{4.24}$$

Equation (4.24) is the condition for critical angle θ_c, and (4.24) can be put into another form as follows:

$$\begin{aligned} \sin \theta_c &= \mathrm{Re}\{N\} = n \\ \mathrm{Im}\{N\} &= k = 0 \, . \end{aligned} \tag{4.25}$$

The first equation is the familiar condition for critical angle, whereas the second states that the medium is nonabsorbing. Equation (4.25) is the basis of many refractometers, such as Abbe-refractometer, which have been constructed for the assessment of the real refractive index of various transparent liquids. When $n < \sin\theta_c$, we observe from (4.4) and (4.5) that $R_s = 1$ and $R_p = 1$, because $(n_2 - \sin 2\theta)^{1/2}$ is a purely imaginary number. However, from formulas (4.1), (4.2) and (4.3) we can find that the phase angles are different for s- and p-polarized light (remember that we are dealing the case $k = 0$). Although our primary interest is reflection of light at this point it is worth to consider also the transmission coefficients of the s-and p-polarized light. Such an idea seems a little bit odd since we anyhow are dealing with the total reflection. However, the reason for the treatment of transmission becomes clear after a moment. The electric field transmission coefficients for nonabsorbing medium are [11]

$$t_s = \frac{2\cos\theta}{\cos\theta + \sqrt{n^2 - \sin^2\theta}}$$

$$t_p = \frac{2\cos\theta}{n\cos\theta + \sqrt{n^2 - \sin^2\theta}} . \tag{4.26}$$

If the square root in (4.26) is purely imaginary, that is to say the angle of incidence is higher than the critical angle but less than $\pi/2$, we find out that in the case of the total reflection surpisignly neither t_s or t_p is zero. Due to this fact we study next the transmitted wave according to the notations of Fig. 4.4. The space-dependent part of the transmitted wave can be expressed as

$$E = E_0 \exp\{-\mathrm{i}(Az\cos\theta_T + Ax\sin\theta_T)\} , \tag{4.27}$$

where A is the wave number of the transmitted wave and θ_T is the angle of refraction. According to the Snell law, and in the case of angle of incidence that exceeds the critical angle of total reflection, we can write an equation that violates the rules of trigonometry of real numbers

$$\sin\theta_T = \frac{\sin\theta_i}{n} > 1 , \tag{4.28}$$

where θ_i is the angle of incidence. If we allow complex angle of refraction then (4.28) is acceptable (see Appendix B). Assuming the validity of (4.28) for total reflection we have to require that

$$\cos\theta_T = \sqrt{1 - \sin^2\theta_T} = \mathrm{i}C . \tag{4.29}$$

where C is a real constant that can be negative or positive. On physical grounds only the negative value of the constant is reasonable. Positive C would lead to the amplification of the transmitted field, which is an absurd situation. Substitution of the information of (4.29) into (4.27) means that

$$E = E_0 \exp(-ACz) \exp\left(-\mathrm{i}Ax(1 + C^2)^{1/2}\right) . \tag{4.30}$$

It is obvious from (4.30) that the wave is evanescent and decays in the optically less dense medium rapidly as z increases. Due to the exponential decay of the amplitude of the evanescent wave energy can not be transported away from the interface. An alternative treatment of the evanescent wave can be based on the analogy of a particle trapped on a finite step potential well. Then the solution of the Schrödinger equation of the particle leads to a standing wave solution inside the well and exponential decay outside the well. In the case of electric field we then assume that there is a probability for the wave to penetrate to the optically thicker medium.

Evanescent wave decaying can be characterized using the concept of penetration depth (d_p) as devised by Harrick [66]

$$E = E_0 \exp(-z/d_p) , \tag{4.31}$$

where E_0 is the electric field at the interfeace. It is assumed that the penetration depth of the evanescent wave in the rearer medium corresponds to the case $E = E_0 \exp(-1)$. Then it holds that

$$d_p = \frac{\lambda}{2\pi n_{\text{prism}} \sqrt{\sin^2 \theta - n^2}} , \tag{4.32}$$

where λ is the wavelength of light in vacuum. Obviously at critical angle the penetration depth is infinite for nonabsorbing liquid. The x-dependence of the wave in (4.30) indicates wave propagation parallel to the interface and hence energy flow in that direction.

If we in turn consider the interaction of the incoming and reflected light fields it can be shown that in the z-direction in the denser medium a standing wave pattern appears. This standing wave pattern can be understood also as a static interference grating. The standing and evanescent waves are illustrated in Fig. 4.5. The above discussion is based on the concept of infinite plane wave. Certainly the finite extent of the real light beam brings complexity in the treatment of the total reflection and corresponding evanescent wave. Reflectivity of a Gaussian (laser) beam near critical angle of absorbing media has been studied in order to estimate the refractive index of the absorbing medium using the first and second derivatives of the reflectance [67].

4.5 Attenuated Total Reflection from Absorbing Liquids

Apparently in the case of absorbing liquid ($k > 0$) the concept of critical angle looses its meaning. This can be seen from the curves of Figs. 4.3b and c, which are drawn for an absorbing medium. The reflectance curves both for s- and p-polarized light have no sharp edges like in the case of nonabsorbing liquids at the location of the critical angle. This indicates e.g. that we can expect that the conventional Abbe-refractometers would give erroneous results for the

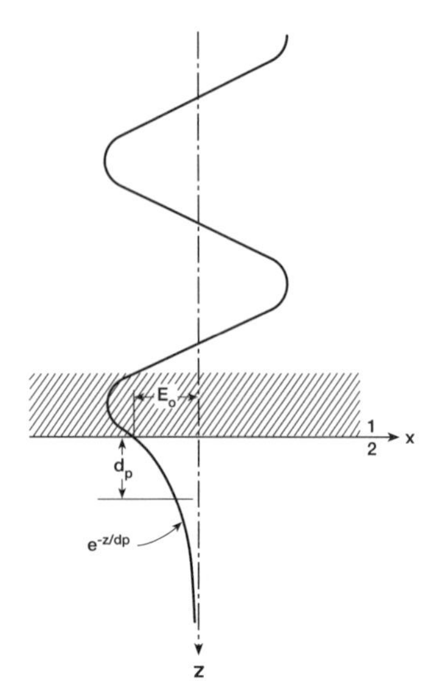

Fig. 4.5. Generation of evanescent wave and a standing wave pattern in total reflection of light. The amplitude of evanescent wave is strongly attenuated in positive z-direction. The depth of penetration of the evanescent wave is d_p

refractive index of the absorbing liquid or they fail to distinguish any critical angle of optically thick liquids. Fortunately inspection of the derivatives of the reflectance of absorbing liquid reveals that there is a turning point on the reflectance curve, which corresponds to the location of the critical angle that can be understood as an apparent critical angle. The location of the apparent critical angle depends, in addition to the extinction coefficient of the liquid, also on the polarization state of the light. This has been demonstrated e.g. by Song et al. [68]. Most convenient way to find the special point of the reflectance curve is based on numerical derivation of the reflectance curve. The apparent critical angle is the one, which fulfills the relations

$$\frac{dR(\theta)}{d\theta} = M$$
$$\frac{d^2R(\theta)}{d\theta^2} = 0\,, \tag{4.33}$$

where M is the maximum value of the first derivative and the reflectance is taken either for s- or p- polarized light. Derivative curves for s- and p-polarized light are illustrated in Fig. 4.6.

In the case of absorbing or turbid liquid some of the energy of the evanescent wave is lost in the liquid due to dissipation or scattering processes. The first mentioned process can be seen in Fig. 4.3b from the reflectance curves which yield always a value less than unity beyond the "invisible" critical angle. Due to the light attenuation by the evanescent field the corresponding

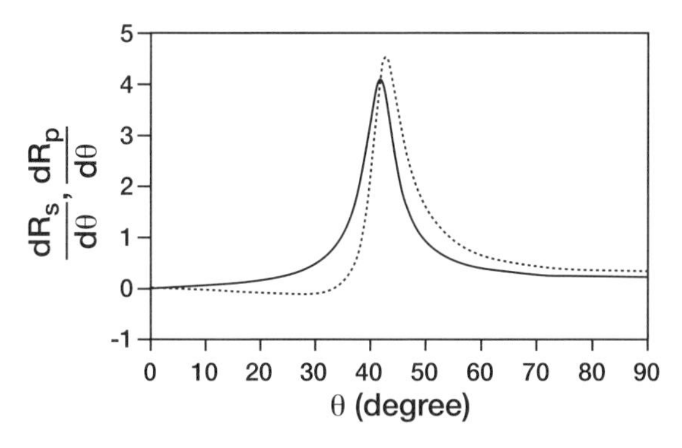

Fig. 4.6. Examples of derivative curves of reflectances for s-and p-polarized light. The medium is absorbing light

total reflection is now termed as attenuated total reflection (ATR). ATR of liquids is nowadays a well-known and sensitive method for the recording of spectra from absorbing liquids and other substances such as powder [52].

The penetration depth in the case of weakly absorbing liquids has been an interesting topic, which we will deal with in more details in next chapter, and there are quite many studies since the seminal study of Harrick [66]. Ekgasit has considered the upper limit of weak absorption [69]. Usually absorption is considered to be weak when $k \ll 0.1$. So far we have been concerned about how to find the real refractive index using the concept of critical angle. Since there are problems in such a procedure, especially in context of absorbing liquids, other more suitable methods must be found. One method is described in the following section.

4.6 Fresnel's Formulas and Data Optimization

The determination of refractive index by reflection using the critical angle method is based on the visual or electronic detection of one specific angle, i.e. the critical angle. Unfortunately, the method becomes inaccurate when the sample exhibits absorption characteristics. Better results are achieved when the whole reflectance curve as a function of the angle of incidence is involved; the R-curve is fitted to a proper theory yielding the complex refractive index. The existence of "extra" data points (more than two data points) leads the data analysis to a least square problem. Let us assume some phenomenon and a corresponding model to describe it. The experimental data y_j, associated with the phenomenon, may be used to estimate the parameter u of the model. Mathematically this can be expressed in a vector form

$$y_j = f(u, t^j) + \tau_j \ , \qquad (4.34)$$

where vector t is a variable of the experimental data, τ_j is an error related to the measurement j and f is the model, i.e. some nonlinear function. The parameter u of the model is obtained if the expression

$$S = \min_u \sum_{j=1}^{m} \left[y_j - f(u, t^j)\right]^2 \tag{4.35}$$

is minimized. Applying the equation to the reflectance measurements, it becomes desirable to replace y_j and $f(u, t^j)$ by the experimental reflectance $R_{\mathrm{exp,j}}$ and the theoretical Fresnel's reflectivity $R_j(N, \theta_j)$. Here the complex refractive index N corresponds to the parameter u of the model. For minimization general optimization procedure e.g. Levenberg–Marquardt method using Minerr function of Mathcad may be used [70]. We remark that during the measurements we gain reflectance for a discrete set of angle variable θ_j, and complex refractive index is obtained for fixed wavelength. The polarization is usually chosen so that light is either s- or p-polarized. Fig. 4.7 shows examples of measured reflectance of s-polarized light from transparent water-ethanol and absorbing water-lignin solutions at wavelengths 590 nm and 280 nm, respectively [71]. The solid line describes the best fit curve obtained from Fresnel's formula (4.1).

4.7 Reflectance from Nonlinear Liquids

The critical angle of total reflection from Kerr liquid depends on the light intensity and a bistability in the reflectance has been theoretically predicted [72] and experimentally shown [73]. This bistability is due to the interaction of the evanescent wave with the optically nonlinear liquid.

The theory of total permittivity of Kerr liquid and a liquid obeying nonlinear optical effective medium model is practically speaking similar, therefore, here we deal only with the case of nanoparticles embedded in a liquid matrix. Suppose that the total effective permittivity obeys the relation

$$\epsilon_{\mathrm{eff}}^{\mathrm{tot}} = \epsilon_{\mathrm{eff}} + \chi_{\mathrm{eff}}^{(3)} I \,, \tag{4.36}$$

where the nonlinear effective permittivity is relatively small compared with the linear one. Then the total effective reflectances for s- and p-polarized light can be expressed as follows:

$$R_{\mathrm{eff,s}}^{\mathrm{tot}} = \left| \frac{\cos\theta - (\epsilon_{\mathrm{eff,r}} - \sin^2\theta)^{1/2}}{\cos\theta + (\epsilon_{\mathrm{eff,r}} - \sin^2\theta)^{1/2}} \right|^2 \tag{4.37}$$

$$R_{\mathrm{eff,p}}^{\mathrm{tot}} = \left| \frac{\epsilon_{\mathrm{eff,r}}\cos\theta - (\epsilon_{\mathrm{eff,r}} - \sin^2\theta)^{1/2}}{\epsilon_{\mathrm{eff,r}}\cos\theta + (\epsilon_{\mathrm{eff,r}} - \sin^2\theta)^{1/2}} \right|^2 , \tag{4.38}$$

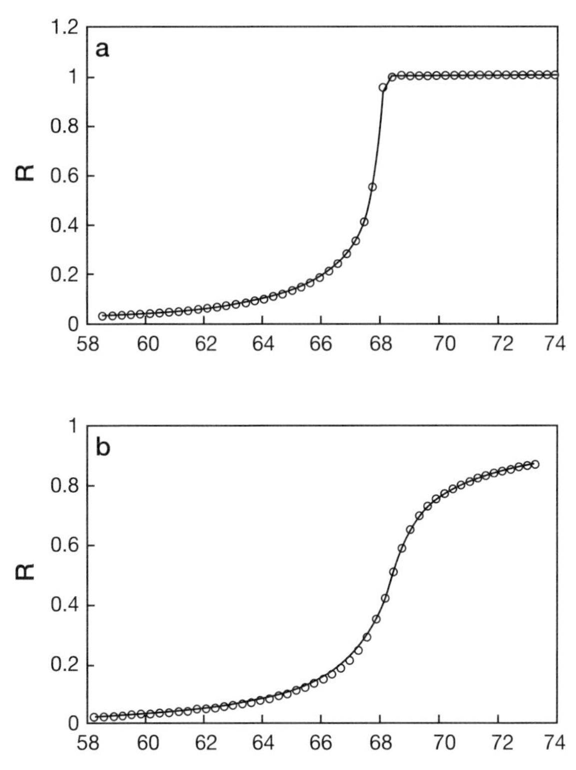

Fig. 4.7. Experimental (o) and theoretical (solid line) reflectances of (**a**) water-ethanol and (**b**) water-ligning solutions at wavelengths 590 nm and 280 nm, respectively

where $\epsilon_{\rm eff,r} = \epsilon_{\rm eff}^{\rm tot}/\epsilon_{\rm prism}$. We next consider the case of ATR-configuration using a fixed light wavelength. Usually the light source is a laser, which provides an intense light beam. In Fig. 4.8 are shown reflectances for a nonlinear Maxwell Garnett two-phase system for different fill fractions and constant light intensity. The calculations are based on the use of (3.9), (3.36), (3.37), (3.43), and (4.36)–(4.38). From Fig. 4.8 we can conclude, because the nanospheres are assumed to absorb light, that the position of the critical angle of the total reflection is obscure also in the case of optically nonlinear liquid. The derivation of the reflectance curve is also possible here in order to find the angle for maximum reflectance. However, the method of Sect. 4.6 is better suited for the determination of the effective complex refractive index of the liquid containing nanospheres, optically either linear or nonlinear.

In Figs. 4.9a and b are shown the difference curves for the total nonlinear and linear reflectance corresponding to the data of Fig. 4.8. It is interesting that the sign of the difference depends on the fill fraction at higher angles of incidence. In Figs. 4.10 and 4.11 are shown data using the same parameters

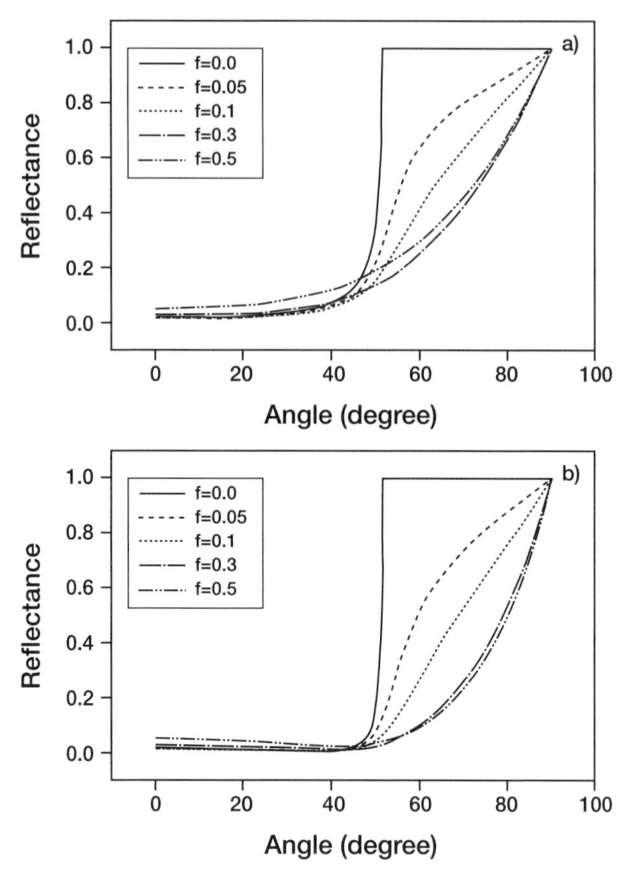

Fig. 4.8. Reflectance curves (**a**) s-polarization and (**b**) p-polarization as afunction of angle of incidence and fill fraction. The wavelenght of light matches the resonace frequency of a Maxwell–Garnett system

as in Figs. 4.8 and 4.9, but using an angular frequency of incident light that is higher than the resonance angular frequency. If we compare Figs. 4.8 and 4.10, we can observe a drastic change. Indeed, the reflectance curves intersect the solid line (the solid line indicates no absorption) at a relatively high reflectance value. The curves in Fig. 4.10 also differ from each other relatively strongly as a function of fill fraction. This means stronger sensitivity of reflectance against nonlinearity and also over to a broader angle range than in Fig. 4.8. Nevertheless, the sensitivity of reflectance against nonlinearity is stronger in the vicinity of the resonance angular frequency. This can be observed by comparing data in Figs. 4.9 and 4.11. The maxima of the difference curves in Fig. 4.11 are roughly twice as large as in those in Fig. 4.9. In addition, in Fig. 4.11 the difference is virtually always negative.

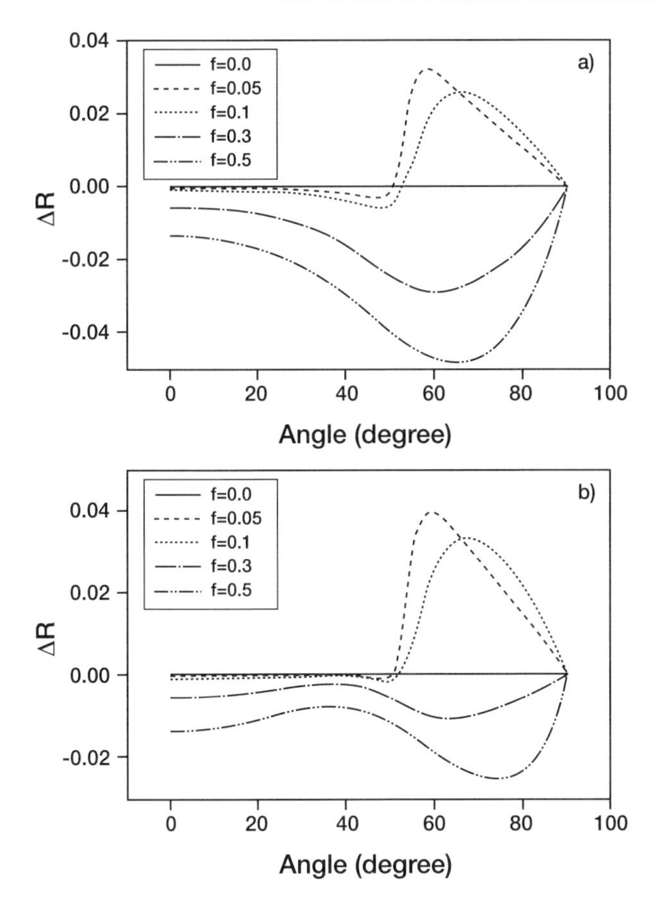

Fig. 4.9. Difference of nonlinear and linear reflectances. Curves were calculated using data of Fig. 4.8

4.8 Fresnel's Formulas and Scattering

Unfortunately, the Fresnel's fomulas (4.4) and (4.5) are usually not valid for light scattering liquids. Meeten and North [47] observed that the reflectance from aqueous suspension, containing polystyrene particles, could be an oscillating function thus departing greatly from the reflectance curve suggested by the Fresnel theory. This is demonstrated in Fig. 4.12. According to Meeten and North the reason for the oscillation could result from enhanced scattering at certain angles by constructive interference. Peiponen et al. [74] observed in their investigations on concentric spherical plastic pigment particles in water (slurry) also oscillation of the reflectance, which depends on the wavelength of the light. They found that it is possible to optimize the wavelenght of light so that best fit with the Fresnel's theory can be found. However, care has to be taken in the interpretation of the complex refractive index calculated

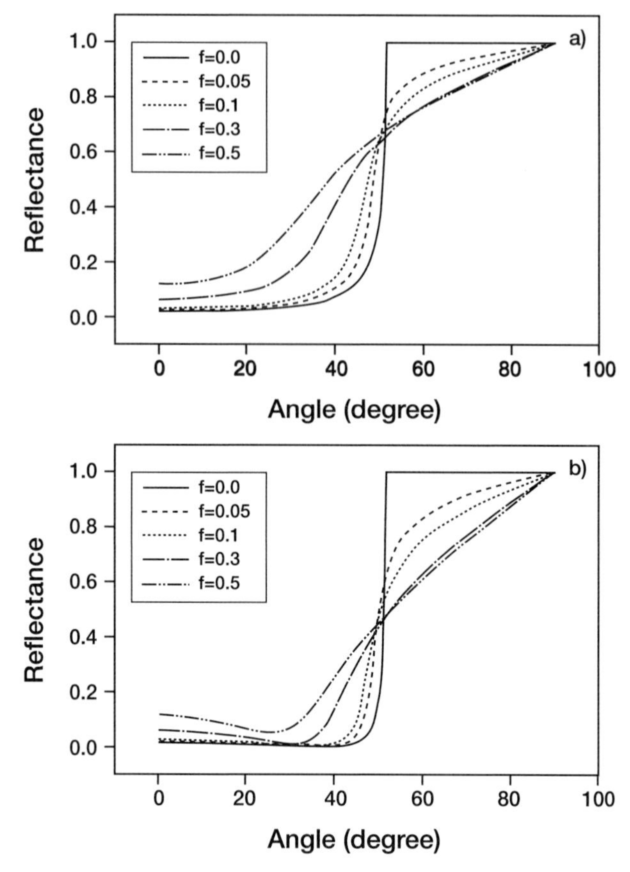

Fig. 4.10. Reflectance curves. (**a**) s-polarization and (**b**) p-polarization. The wavelenght is different to the resonance angular frequency of a Maxwell–Garnett system. The parameters of calculations were the same as in Fig. 4.8

from the reflectance of a turbid fluid, since usually the reflectance signal can carry photons, which are related to surface and/or volume scattering from the sample. In addition the forward-scattering formula (3.56) has to be considered as an approximation in order to find consistency between the reflectance obtained by Fresnel's theory and measurements.

An example describing of the non-Fresnel feature of the reflectance from a turbid liquid is presented in Fig. 4.13a, where the solid line describes the Fresnel theory and the dots are experimental data from milk. An empiric formula for s-polarized light effective reflectance from turbid milk was proposed by Räty and Peiponen [75], which is based on the exploitation of (4.35).

$$R_{\mathrm{eff,s}} = R_s + R_s^{1.5}(1 - R_s) \ , \tag{4.39}$$

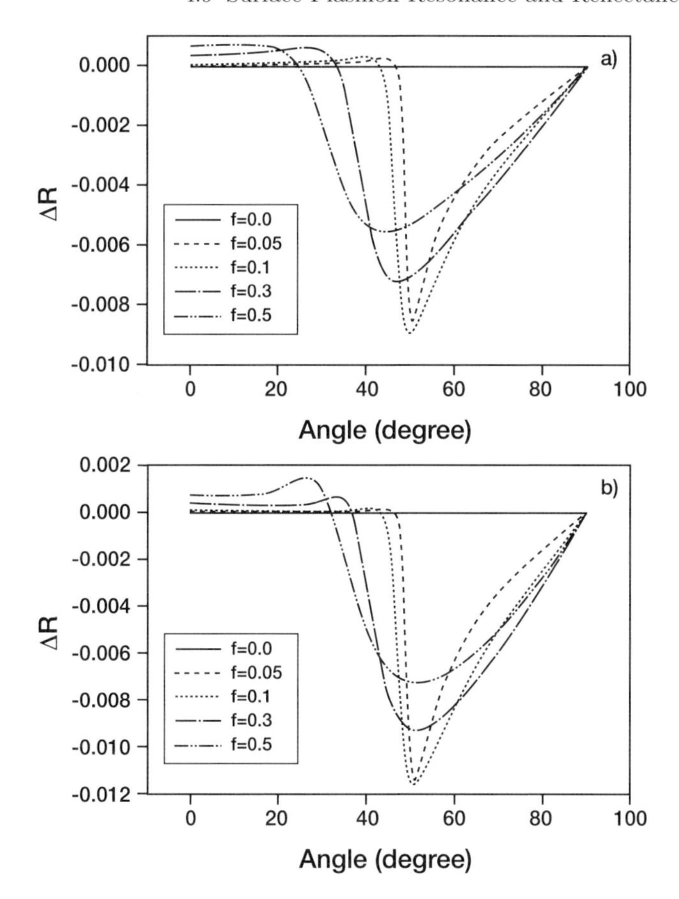

Fig. 4.11. Difference between nonlinear and linear reflectances corresponding data of Fig. 4.10

where R_s is obtained from (4.4). Better agreement between measurement and theory can now be observed in Fig. 4.13 b), where the model of (4.39) has been employed.

4.9 Surface Plasmon Resonance and Reflectance

A plasmon is a collective oscillation of electrons in metals and the rigorous description of the phenomenon is based on quantum mechanics [9]. Surface plasmons exist in the boundary of the metal and they can be produced in some cases using an external electric field. Let us first consider the volume plasmon using the concepts of classical electromagnetism. The relative permittivity of an insulator can be qualitatively described with the aid of (3.9). In the case of metals there is no restoring force that would keep conduction electrons

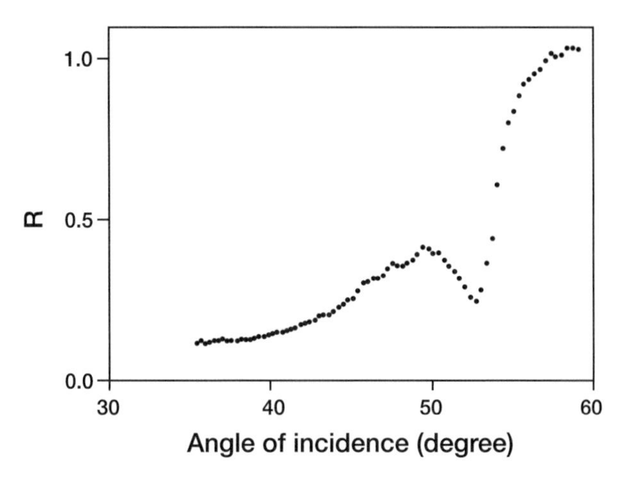

Fig. 4.12. Oscillation of reflectance as a function of angle of incidence due to Latex particles in liquid phase

confined in the vicinity of the nucleus. In other words the spring coefficient $b = 0$. This in turn means that also $\omega_0 = 0$. If we substitute this information into (3.9) we get the relative complex permittivity of a metal (ϵ_{mr}). The resulting relations are called Drude dispersion formulas, and they are

$$\begin{aligned} \mathrm{Re}\{\epsilon_{mr}\} &= 1 - \frac{\rho e^2}{m\epsilon_0} \frac{1}{\omega^2 + \Gamma^2} \\ \mathrm{Im}\{\epsilon_{mr}\} &= \frac{\rho e^2}{m\epsilon_0} \frac{\Gamma}{\omega(\omega^2 + \Gamma^2)} \ . \end{aligned} \tag{4.40}$$

Plasma frequency (ω_p) of the metal is defined by the relation that $\mathrm{Re}\{\epsilon_{mr}\} = 0$. This happens when $\omega \gg \Gamma$ and the plasma frequency is given by the definition

$$\omega_p^2 = \frac{\rho e^2}{m\epsilon_0} \ . \tag{4.41}$$

According to (4.40) and (4.41) we can approximate at high frequencies

$$\begin{aligned} \mathrm{Re}\{\epsilon_{mr}\} &\cong 1 - \frac{\omega_p^2}{\omega^2} \\ \mathrm{Im}\{\epsilon_{mr}\} &<< 1 \ . \end{aligned} \tag{4.42}$$

Furthermore, it follows from (4.4) and (4.5) that the reflectance has a sudden decrease at the plasma frequency, which is usually at the UV–VIS range for real metals. The volume plasmons, which are fluctuations in charge density, can be observed using a beam of electrons for exitation.

Here we are intrested in surface plasmon resonance (SPR), which can be exited using a light beam. As concerns the light sensing applications

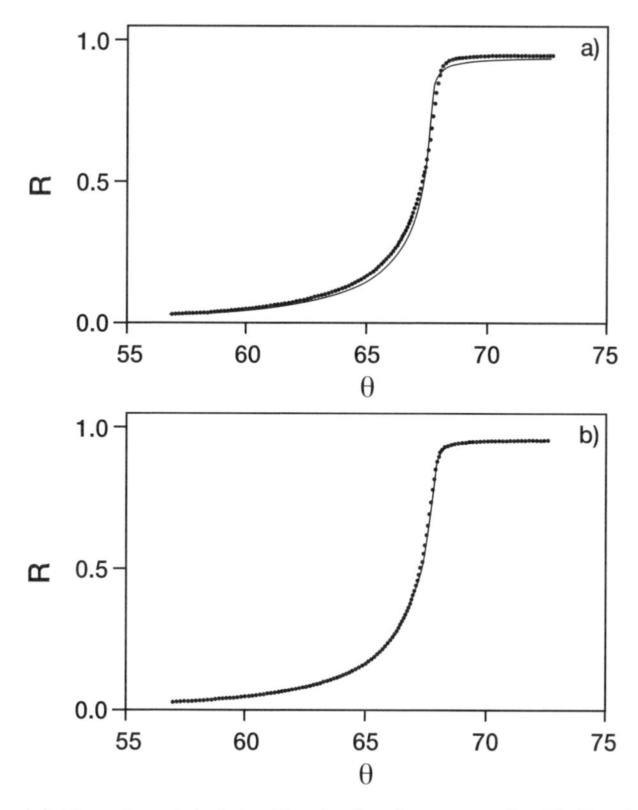

Fig. 4.13. (**a**) Experimental data (dots) of reflectance of milk fitted to Fresnel theory (solid line), and (**b**) fitting by modified Fresnel equation (4.39). The probe wavelength was 590 nm

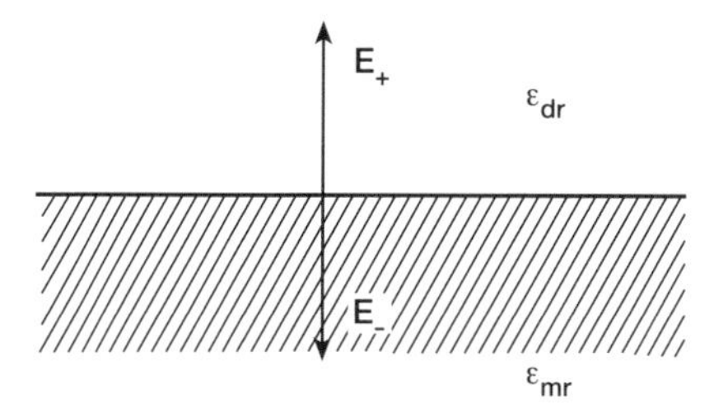

Fig. 4.14. Opposite electric fields at the interface of a metal and an insulator

and a prism reflectometer related to surface plasma waves the studies of Kretschmann and Räther [76] and Kretschmann [77] have much importance. A thorough description of theory and applications of SPR can be found from the book [78], and a book chapter written by Räther [79] and also from a review written by Homola et al. [80]. Consider the system of Fig. 4.14, then the divergence of the electric flux ($\vec{D}$), is

$$\nabla \cdot \vec{D} = \epsilon_{dr} E_+ - \epsilon_{mr} E_- , \tag{4.43}$$

where ϵ_{dr} is the relative permittivity of the dielectric medium shown in Fig. 4.14. In the absence of external charges, the electric field arises only from polarization charges on the boundary then by the symmetry it holds that $E_+ = -E_-$. Then it follows from (4.43)

$$\epsilon_{mr} = -\epsilon_{dr} . \tag{4.44}$$

If we substitute the approximation (4.42) of the real part of the permittivity into the left hand-side of (4.44) we find out that the surface plasmon frequency is

$$\omega_S = \frac{\omega_p}{\sqrt{\epsilon_{dr} + 1}} . \tag{4.45}$$

In the special case of small spherical metal particles [8] in a dielectric matrix the surface plasma frequency is equal to $\omega_p/(2\epsilon_{mr} + 1)^{1/2}$. It follows from Maxwell's equations that when a light beam has a p-polarized light component only that component can generate a surface plasma wave [81]. The oscillation of surface charge fluctuations cannot ordinarily be exited by light. Otto [82, 83] and Kretschmann and Räther [76] proposed the exploitation of a prism and a thin metallic film for surface plasma wave generation. Then the observation of reflectance in ATR-mode provides information about a resonance. That is to say, at specific angle of incidence the reflectance has a dip due to the fact that light beam is coupled most effectively into the metal film. The resonance angle is larger than the critical angle and it depends on the complex permittivitty of the metal film, the optical properties of the liquid (or gas) to be studied, and the refractive index of the prism. Now the theory is somewhat more complicated than in the case of bulk metal above. The reflectance can be derived by inspection of multiple light reflection in an ambient-film-substrate system [84]. The expression of the reflectance is as follows:

$$R_p(\theta) = \left| \frac{r_{pm}(\theta) + r_{ml}(\theta) \exp[2\mathrm{i}A_z(\theta)d]}{1 + r_{pm}(\theta) r_{ml}(\theta) \exp[2\mathrm{i}A_z(\theta)d]} \right|^2 , \tag{4.46}$$

where r_{pm} is the electric field reflectance at prism-metal film interface, r_{ml} is the corresponding reflectance at the metal-liquid interface, d is the thickness (typically around 50 nm) of the metal film, and A_z is scalar component of wave vector normal to the metal film surface. The electric field reflectances are

$$r_{pm} = \frac{\frac{A_{\mathrm{z,prism}}}{\epsilon_{\mathrm{prism,r}}} - \frac{A_{\mathrm{zm}}}{\epsilon_{mr}}}{\frac{A_{\mathrm{z,prism}}}{\epsilon_{\mathrm{prism,r}}} + \frac{A_{\mathrm{zm}}}{\epsilon_{mr}}} \tag{4.47}$$

$$r_{ml} = \frac{\frac{A_{\mathrm{zm}}}{\epsilon_{mr}} - \frac{A_{\mathrm{z,lig}}}{\epsilon_{\mathrm{lig,r}}}}{\frac{A_{\mathrm{zm}}}{\epsilon_{mr}} + \frac{A_{\mathrm{z,lig}}}{\epsilon_{\mathrm{lig,r}}}} , \tag{4.48}$$

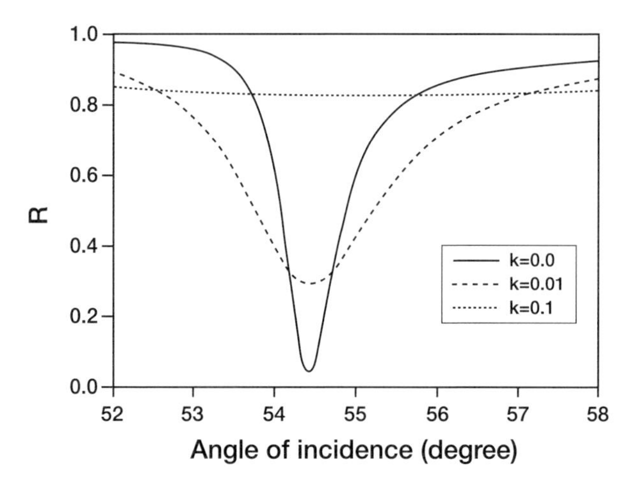

Fig. 4.15. Dip of reflectance due to surface-plasmon-resonance. The level and the width of the dip depend on the extinction coefficient k of the liquid

where $\epsilon_{\mathrm{prism,r}}$ is the relative permittivity of the prism, ϵ_{mr} is the complex relative permittivity of metal, and $\epsilon_{\mathrm{liq,r}}$ is the complex relative permittivity of the liquid. The wavenumber is defined as follows:

$$A_{zj} = \left[\epsilon_{jr}\left(\frac{\omega}{c}\right)^2 - A_x^2\right]^{1/2} , \tag{4.49}$$

where

$$A_x = n_{\mathrm{prism}}\left(\frac{\omega}{c}\right)\sin\theta . \tag{4.50}$$

In Fig. 4.15 is shown a reflectance curve for water ($k = 0$) calculated using (4.46)–(4.50) for the case of a 50 nm thick silver film. In the calculation the prism refractive index is taken to be equalt to FK11 glass for HeNe laser wavelength 632.8 nm. The minimum of the reflectance curve in Fig. 4.15 corresponds to the surface plasmon resonance, which can be shown with the aid of the theory of electromagnetism, to be obtained at an angle θ_{sp} and it holds.

$$n_{\mathrm{prism}}\sin\theta_{sp} = \mathrm{Re}\left\{\sqrt{\frac{\epsilon_{mr}\epsilon_{\mathrm{lig,r}}}{\epsilon_{mr}+\epsilon_{\mathrm{lig,r}}}}\right\} . \tag{4.51}$$

This relation is widely used for the assessment of the refractive index of nonabsorbing liquids since the permittivity of the metal is usually known and the angle of surface plasma resonance is experimentally detected. SPR signal can be detected if the real part of the permittivity of the metal film is negative and smaller than the negative value of the squared refractive index of the liquid. In the case of absorbing liquid the half width and the depth of the dip depend on the extinction coefficient of the liquid. This is also demonstrated in Fig. 4.15. At the vicinity of the dip the reflectance can be approximated using a Lorentzian line model [85, 86], which makes it possible to estimate e.g. the permittivity of the metal film and its thickness, and their uncertainities [87]. Kano and Kawata [88] investigated enhancement for absorption-sensitivity of surface-plasmon sensor by optimizing the metal film thickness. Unfortunately, the thin metallic film is usually subject to wear in severe industrial measurement environments, which may cause erroneous measurement results.

Surface plasmon resonance (SPR) for material sensing has turned out to be a very sensitive technique to detect small changes of the refractive index of gaseous [89] and liquid phase [80]. Nowadays there are various measurement techniques, which employ e.g. a grating configuration instead of the prism, in the detection of physico-chemical changes in media based on SPR sensing. Thus SPR has proved to be a valuable tool e.g. in the analysis of dynamic biological interactions [2, 3]. A popular device, which is based on Kretschmann's configuration, exploits a flow cell [90] that introduces analyte solution, which passes the thin metal film of the prism. The metal film is e.g. polymer-coated and adsorption of proteins to the polymer film is monitored by detection of the time-dependent SPR. Figure 4.16 illustrates this type of monitoring of kinetics of adsorption.

Refractive index and hence information about concentration of constituents of "ill-behaved" industrial liquids such as turbid pulping liquor [91], pigment slurries [92] or milk [93] can be obtained by detection of surface plasmon resonance from these liquids but usually in laboratory conditions.

Matsubara et al. [94] introduced a device, which makes use of a convergent light beam. Then there is no need to rotate the prism. The last mentioned technique has been exploited also in commercial devices [80]. The limitation of such a method is usually that only a relatively narrow refractive index range can usually be covered. Johansen et al [95] suggested that it is (numerically) possible to achieve a resolution better than 10^{-9} refractive index units (RIU).

Another technique is based also on the Kretschmann's configuration but keeping the angle of incidence fixed and instead using a wide band white light source for SPR generation [96]. In that mode only a certain wavelength of the used spectral range will couple to a surface plasma wave related to a specific constituent of the analyte. Multi-wavelength technique for assessment of optical constants of liquids has been proposed by Yee [97–99], and by Lavers and Wilkinson [100]. Räty et al [101] introduced a multifunction reflectometer which makes it possible to attain wavelength-dependent complex refractive index of liquids in SPR measurement mode. Furthermore Saarinen et al [102] developed an analysis method, which allows the calculations of the complex refractive index of a liquid using SPR-reflectance obtained at a fixed angle of light incidence but scanning the wavelength of the light. In that scheme the unknown wavelength-dependent complex permittivity of the liquid is obtained from (4.46)–(4.48) as follows:

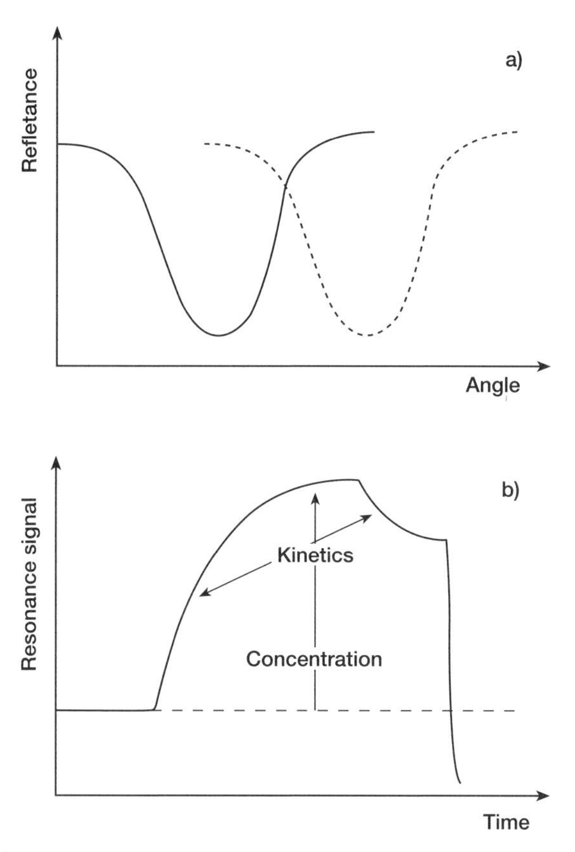

Fig. 4.16. (**a**) The change of the dip as a result in concentration change of the analyte, and (**b**) temporal change of the signal for a fixed angle of incidence

$$\epsilon_{\mathrm{lig,r}} = \epsilon_{\mathrm{prism,r}} \frac{A_{\mathrm{z,lig}}(1-r_{pm})(1-r_{ml})}{A_{\mathrm{z,prism}}(1+r_{pm})(1+r_{ml})} . \tag{4.52}$$

After some algebra (see the detailed derivation presented in [102]) and utilizing (4.49), equation (4.52) can be put to into a more practical form

$$\epsilon_{\mathrm{lig,r}} = \frac{1}{2}[C + (C^2 - 4C\epsilon_{\mathrm{prism,r}} \sin^2\theta)^{1/2}] , \tag{4.53}$$

where

$$C = \frac{2\pi}{\lambda} \left[\frac{\epsilon_{\mathrm{prism,r}}(1-r_{pm})(1-r_{ml})}{A_{\mathrm{z,prism}}(1+r_{pm})(1+r_{ml})} \right] . \tag{4.54}$$

The method is based on phase retrieval procedure, which means that the phase of the complex reflectance appearing inside the modulus, on right-hand side of (4.46), is calculated from the wavelength-dependent R_p. An example of such a calculation is postponed in the section 6.4 where maximum entropy method is introduced.

The choice of metal and wavelength for surface-plasmon resonance sensors has been also investigated [103]. Due to the high sensitivity of SPR sensor on refractive index of media the sensor has also strong sensitivity on temperature change, which has to be properly taken into account [104].

Microscopy [105–109], related to surface plasma wave generation, provides another tool for materials research.

5 Probe Window Contamination and Reflectance

In practical metrology, especially in industrial environments, the measurement conditions are usually hostile. If we consider for instance the quality assessment of process liquids in process industry such as paper mill the contamination of the probing face of prism reflectometer is a drawback. In other words different constituents of the liquid can be adsorbed on the probe face. This in turn causes change in the reflectance signal either because of the change of the complex refractive index or because of light scattering in the vicinity of interface between the liquid and the probe face. In this chapter we briefly describe simple models that may be used for evaluation of the growth of contamination on the probe window. Then we are talking about adsorption. In addition we consider a totally contaminated layer and related depth profiling.

5.1 Adsorption

Reflectometry can be exploited as a tool for adsorption studies. For instance Dijt et al. [50] have applied reflectometry for investigation of adsorption from dilute solutions. Unfortunately, process liquids are usually not dilute and there appears quite often a dynamic growth of the contaminated layer on the prism face. That is to say there are locations on the probe window that are clean while some are contaminated. After some saturation time the whole probe face of the prism reflectometer is contaminated. Then one can ask "is it anymore possible to apply e.g. Fresnel's formulas, and is it possible to monitor the true optical properties of the liquid?" These questions are usually quite difficult to answer. However, there are some simplifying models that can be used for assessment of the optical constants of liquids despite of the contamination of the probe face. In principle the Bruggeman model of Sect. 3.4 can be exploited, especially for geometric factor $g = 1/2$, to the estimation of the reflectance related to the probe window contamination in the nanoscale. Then the contaminated and noncontaminated domains form islands in the presence of percolation. Most often the contamination is macroscopical in the plane of the probe face, whereas it can be microscopical in the direction normal to the face. Here we present a simple model of probe window contamination, which is based on the idea of Mäkinen et al. [110]. Suppose that the contamination particles are adhering only onto clean surface of the probe window and the thickness of the contamination layer is not increasing, but the clean surface area is diminishing as a function of time while contact with a (turbid) liquid. Then the reflectance should not change after the whole probe window is contaminated. Next we assume that the surface of the probe window is divided into infinitesimal areas each unit area is either clean or contaminated. The adsorption of contamination particles is assumed to be a random process. The probability to be contaminated in differential time dt is $Prdt$, where Pr is the probability density. Furthermore, we assume that each unit area is optically similar and at the instant $t = 0$ the probe window is clean. Under these assumptions the total reflectance from a probe window is a geometric sum of reflectances from clean (R_{cl}) and contaminated (R_{co}) areas as follows:

$$R(\theta) = S_{\mathrm{cl}} R_{\mathrm{cl}}(\theta) + S_{\mathrm{co}} R_{\mathrm{co}}(\theta) \, , \tag{5.1}$$

where S_{cl} and S_{co} are surface fractions of clean and contaminated areas, thus it holds that

$$S_{\mathrm{cl}} + S_{\mathrm{co}} = 1 \, . \tag{5.2}$$

When the contamination particles are adhering in a random manner onto the clean surface the clean area is decreasing proportionally to

$$-dS_{\mathrm{cl}} = S_{\mathrm{cl}} Prdt \, . \tag{5.3}$$

The minus sign indicates decrease of the clean area. Equation (5.3) is a first-order differential equation, which has the well-known solution

$$S_{\mathrm{cl}} = e^{-Prt} \, . \tag{5.4}$$

Substitution of (5.2) and (5.4) into (5.1) yields

$$R(\theta, t) = R_{\mathrm{cl}}(\theta) e^{-Prt} + R_{\mathrm{co}}(\theta)(1 - e^{-Prt}) \, . \tag{5.5}$$

At the beginning ($t = 0$) the reflectance is equal to $R_{t=0} = R_{\mathrm{cl}}$ and when $t = \infty$ the reflectance is equal to $R_{t=\infty} = R_{\mathrm{co}}$. The last mentioned reflectances are obtained by measurement. Parameters R_{cl} and Pr can be solved by minimizing the least square of the difference of experimental data and (5.5).

5.2 Depth Profiling

In Sect. 4.4 we defined penetration depth (4.32) for nonabsorbing liquid. Next we consider the case of weak absorption in the context of adsorbed layer on the probe window. Here we assume that the whole window area is now contaminated but the concentration of the contaminated layer is changing in the direction perpendicular to the prism face, which is in contact with the weakly absorbing liquid. The pioneer of this subject was Harrick [52]. It follows from (3.30) that we can approximate the transmission of weakly absorbing medium using the Taylor series expansion

$$\frac{I}{I_0} \approx 1 - \alpha d \, . \tag{5.6}$$

By analogy in ATR the reflectance in the case of weak absorption is

$$R = 1 - \alpha d_{\text{eff}} \tag{5.7}$$

and d_{eff} is the effective thickness defined by Harrick [52, 111] so that it corresponds the thickness of contamination film which would give the same absorption for transmission at normal incidence. Then the effective thickness, which takes different mathematical expressions for s- and p-polarized light, respectively, can be obtained. For the case of thin film it holds that

$$\alpha d_{\text{eff}} = \frac{n E_0^2 d}{\cos\theta} \, , \tag{5.8}$$

where d is the film thickness and the amplitude E_0 depends on the polarization of the light. Derivation of (5.8) is based on the assumption that there is no absorption gradient in the contamination layer. This is not usually true, but the absorption coefficient $\alpha = \alpha(z)$. In the case of nonuniform absorption Hirscfeld [112] observed that absorbance is the Laplace transfrom of the absorption coefficient i.e.

$$1 - R(\theta) = \frac{n E_0^2 d}{\cos\theta} \int_0^{\infty} \alpha(z) \exp(-2z/d_p) dz \, . \tag{5.9}$$

The inversion of the integral yields information on $\alpha(z)$ when the angle of incidence is changed. The problem in practical metrology is usually that the reflectance is changing gradually while the angle of incidence is increasing beyond the critical angle. Another problem is the data handling. Nevertheless, in infrared spectroscopy the theory of depth profiling has drawn attention [113–116].

5.3 Matrix Theory for Reflectance of Multilayers

Fresnel equations presented in (4.4) and (4.5) may be directly employed for a system of one plane interface. Here the boundary of the system being formed by two homogenous and linear materials. However, there are optical systems which are constituted of several layers such as interference filters. As a consequence multiple reflection take place. In such a case use of the Fresnel equations is quite awkward.

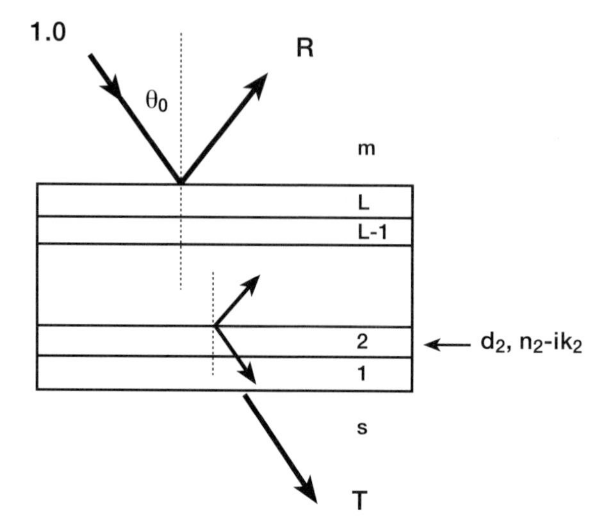

Fig. 5.1. A cross-section of a multilayered media. Each layer is characterized by the thickness and the complex refractive index. For example the parameters for the layer 2 are d_2 and $n_2 - ik_2$

In some cases a contatinant layer may be modelled using the stratified medium theory. Multilayer systems may be analyzed by using a matrix formulation presented by Weinstein [117] and Herpin [118]. The formulation is based on boundary condition at film surfaces derived from Maxwell's equations. The notations applied below follows to that of [119].

Consider a layer system of L individual layers placed on a substrate (s). Thickness of each layer is denoted by d_j and correspondingly the complex refractive index by n_j. Furthermore, the refractive index of semi-infinite medium and substrate are n_m and n_s, respectively (n_s may be complex). Light enters the system from a lossless medium (m) at an arbitrary incident angle of θ_0. In addition, we assume that light is either s- or p-polarized.

The *effective refractive index* of layers as well as of medium and substrate is then given by

$$\eta = \begin{cases} n/\cos\theta & p-polarization \\ n\cos\theta & s-polarization \,. \end{cases} \tag{5.10}$$

The angle θ_j may be determined using Snell's law

$$n_m \sin\theta_0 = n_j \sin\theta_j \,. \tag{5.11}$$

Furthermore, the *effective optical thickness* of a layer j for an angle θ_j is defined by

$$\delta_j' = n_j d_j \cos\theta_j \,. \tag{5.12}$$

The functioning of each layer is governed by a 2×2 matrix M_j. This takes a form of

$$M_j = \begin{pmatrix} \cos\delta_j & \frac{\mathrm{i}}{\eta_j}\sin\delta_j \\ \mathrm{i}\eta_j\sin\delta_j & \cos\delta_j \end{pmatrix} , \tag{5.13}$$

where $\delta_j = (2\pi/\lambda)\delta'_j$. Overall effect of layers is obtained by the multiplication of all matrices

$$M = M_L M_{L-1} \cdots M_1 . \tag{5.14}$$

It can be shown that amplitude reflection coefficient r of such layered system is

$$r = \frac{\eta_m \boldsymbol{E}_m - \boldsymbol{H}_m}{\eta_m \boldsymbol{E}_m + \boldsymbol{H}_m} , \tag{5.15}$$

where $\boldsymbol{E}_m$ and $\boldsymbol{H}_m$ are electric and magnetic vectors, respectively, and are given by

$$\begin{pmatrix} \boldsymbol{E}_m \\ \boldsymbol{H}_m \end{pmatrix} = M \begin{pmatrix} 1 \\ \eta_s \end{pmatrix} . \tag{5.16}$$

If we are also interested in transmittance then we may employ the equation for amplitude transmission coefficient t

$$t = \frac{2\eta_m}{\eta_m \boldsymbol{E}_m + \boldsymbol{H}_m} . \tag{5.17}$$

Finally we get reflectance R and transmittance T as follows

$$R = |r|^2 \tag{5.18}$$

and

$$T = \frac{\eta_s}{\eta_m}|t|^2 . \tag{5.19}$$

6 Wavelength Spectra Analysis

A great deal of our knowledge related to spectroscopic properties, such as dispersion and absorption of light, of media is based on the exploitation of Kramers–Kronig (K–K) dispersion relations [120–122]. The idea of Kramers–Kronig relations for complex electric field reflectance is that the phase of the electric field can be calculated, as a function of wavelength, with the aid of the intensity reflectance, which is measured. Then the complex refractive index is obtained e.g. from (4.1) or (4.2). Probably the most common situation is where the dispersion relations are applied for normal reflectance i.e. reflectance that is measured at normal incidence. Unfortunately, the rigorous derivation of K–K relations is usually neglected in the literature of the field, probably due to some mathematics, which usually involves complex analysis. However, in the derivation some crucial assumptions are made and one has to be convinced that these assumptions are fulfilled in order to apply K-K relations. Therefore we will spend some time in dealing with the assumptions and derivations of the K–K relations. We will also point out that especially in nonlinear optics the K–K relations can be invalid. As an alternative method, which has been pointed out to be more powerful than K–K relations, we deal with maximum entropy model in phase retrieval problems of spectroscopy. We begin the presentation, due to historical reasons, by derivation of the K–K relations of the complex refractive index. Such relations have importance especially in transmission spectroscopy. The results for the complex refractive index are helpful when we derive the K–K relations for the complex reflectance.

6.1 Kramers–Kronig Relations for Complex Refractive Index

The underlying principle of the existence of K–K relations in linear optical spectroscopy is the causality. Indeed, the K–K relations can be considered as a restatement of causality [13, 123, 124].

The assumptions that are crucial for the existence of the K–K relations are: (1) the optical function (complex refractive index, permittivity or reflectance) is an analytic function in the upper-half of complex angular frequency plane, and (2) the optical function has strong enough asymptotic fall-off when the angular frequency tends to infinity. For the sake of simplicity we employ here the classical qualitative model of a Lorentzian complex relative permittivity of an insulator i.e. the model based on (3.8) and (3.9), but introducing a complex angular frequency variable $\Omega = \omega + \mathrm{i}\xi$, which replaces the real angular frequency variable ω

$$\epsilon(\Omega) = 1 + \frac{\rho e^2}{m\epsilon_0} \frac{1}{\omega_0^2 - \Omega^2 - \mathrm{i}\Gamma\Omega} . \tag{6.1}$$

The transfer into complex variable is important, since we obtain the K–K relations using complex contour integration. From (6.1) we can observe that the complex relative permittivity has poles. That is to say singular points at which the permittivity is infinite. The poles are the roots of

$$\omega_0^2 - \Omega^2 - \mathrm{i}\Gamma\Omega = 0 . \tag{6.2}$$

The solutions of (6.2) are as follows:

$$\begin{aligned} \Omega_{p1} &= \frac{-\mathrm{i}\Gamma + \sqrt{4\omega_0^2 - \Gamma^2}}{2} \\ \Omega_{p2} &= \frac{-\mathrm{i}\Gamma - \sqrt{4\omega_0^2 - \Gamma^2}}{2} . \end{aligned} \tag{6.3}$$

From (6.3) we can deduce that irrespect of the sign of $\Gamma^2 - 4\omega_0^2$ the poles are always located in the lower half plane. The Loretzian complex permittivity is a derivable as a function of the complex angular frequency in the upper-half plane, which can be confirmed by inspection of the Cauchy-Riemann equations (see Appendix C). The existence of the complex derivative means, that the complex relative permittivity is an analytic function in the upper-half plane. However, we wish to study the complex refractive index, which was given in (3.10), with the aid of the relative permittivity. From (3.10) we can solve

$$N(\Omega) = \sqrt{\epsilon_r(\Omega)} . \tag{6.4}$$

At first we could imagine that there are problems with the analyticity of the complex refractive index because of the square root of (6.4). From the theory of complex analysis [125] we know that the square root of a complex function is a double-valued function [126], and the square root of (6.4) might have branch points [126] at

$$\epsilon_r(\Omega) = 0 . \tag{6.5}$$

Due to the physical restriction we can accept only the positive square root (6.4). The physical restriction is that while the angular frequency tends to infinity the complex relative permittivity tends to unity. In other words at infinite high energy the absorption of electromagnetic field in a liquid, or generally in any medium, will vanish and the field experiences the medium as if it were the vacuum. Also the complex refractive index tends to unity i.e. $N = 1$, and the value $N = -1$ is not physically reasonable. This means that we have a single-valued function. The branch points of (6.5) are only apparent. The "branch points" of an insulator would correspond to the plasma frequency of (4.41), but because the imaginary part of the relative permittivity is not zero, then this physical constraint denies the existence of the branch points. The deduction was based on he complex refractive index obtained with the aid of Lorentzian complex permittivity. The deduction can be generalized to be independent on the model. Now the square root of (6.4) is derivable and we have

$$\frac{dN(\Omega)}{d\Omega} = \frac{1}{2}[\epsilon_r(\Omega)]^{-1/2}\frac{d\epsilon_r(\Omega)}{d\Omega} \,. \tag{6.6}$$

On the right-side of (6.6) the inverse of the square root is finite in the upper-half plane, due to the lack of branch points in the upper half plane, and the complex derivative of the complex relative permittivity exists as we have already observed. Therefore, the complex refractive index is an analytic function in the upper-half plane.

The asymptotic fall-off of the complex refractive index can be estimated also using the Lorentzian complex relative permittivity. From (6.1) we find out that as $\Omega \to \infty$ then

$$\epsilon_r(\Omega) - 1 \propto= -\frac{1}{\Omega^2} \,. \tag{6.7}$$

Usually the asymptotic fall-off is stated using the real angular frequency and the relations (4.41). At infinite high angular frequency it follows from the binomial expansion of (6.4) and from (6.1) that

$$N(\Omega) - 1 = \sqrt{\epsilon_r(\Omega)} - 1 \cong \frac{1}{2}\left[Re\{\epsilon_r(\Omega)\} + \mathrm{Im}\{\epsilon_r(\Omega)\}\right] + \cdots \,. \tag{6.8}$$

It is obvious from (6.8) that the complex refractive index falls-off proportionally to Ω^{-2}. The last mentioned property is needed for the convergence of an integral. Now we apply the complex contour integration as shown in Fig. 6.1. An artificial pole ω' is introduced on the real axis. The integral that we want is along the real axis extending from minus infinity to plus infinity. For this purpose we integrate the function $(N(\Omega) - 1)/(\Omega - \omega')$, which apparently is singular when $\Omega = \omega'$. In order to avoid the singular nature in integration we consider the Cauchy principal value. The Cauchy principal value is here defined as

$$\begin{aligned} &P\int_{-\infty}^{\infty} \frac{N(\omega)-1}{\omega-\omega'}d\omega \\ &= \lim_{\zeta\to 0, B\to\infty}\left\{\oint_C \frac{N(\omega)-1}{\Omega-\omega'}d\Omega - \int_D \frac{N(\omega)-1}{\Omega-\omega'}d\Omega - \int_A \frac{N(\omega)-1}{\Omega-\omega'}d\Omega\right\} \end{aligned} \tag{6.9}$$

where C is the closed contour, D is the arc of the larger semicircle and A is the small semicircular detour, which avoids the pole on the real axis. The

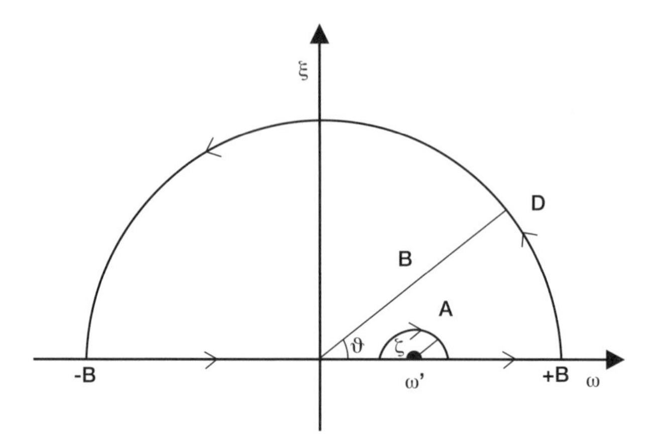

Fig. 6.1. Contours for derivation of Kramers–Kronig relations for complex refractive index

radius of the large semicircle is B and it is allowed to extent to infinity, while the radius of the small ζ tends to zero. The symbol "P" denotes taking of Cauchy principal value i.e. the limiting process, which involves symmetric approach of the pole as ζ tends to zero. Let us consider the first integral on right-hand side of equation (6.9). The integrand is an analytic function on and inside the closed contour. Then according to the Cauchy's intergral theorem it holds that

$$\oint_C \frac{N(\Omega)-1}{\Omega-\omega'}d\Omega = 0\,. \tag{6.10}$$

The limiting processes will not change the result of (6.10). Next we estimate the second integral on the right-hand side of (6.9). The integration is counter clockwise integration along the arc of the large semicircle. The arc of the semicircle is a closed (compact) set [125]. Since the complex refractive index is an analytic function on the closed set it is also uniformly continuous on the arc D. Therefore the complex refractive index will have a maximum value on the arc D

$$M(B) = \max|N(\Omega)-1|;|\Omega| = B, \mathrm{Im}\{\Omega\} \geq 0\,. \tag{6.11}$$

It follows from (6.7), (6.8) and (6.11) that

$$\lim_{B\to\infty} M(B) = 0\,. \tag{6.12}$$

Using the notations of Fig. 6.1, we can estimate

$$\left|\int_D \frac{N(\Omega)-1}{Be^{\mathrm{i}\vartheta}-\omega'}\right| \leq \int_D \frac{|N(\Omega)-1||d\Omega|}{|Be^{\mathrm{i}\vartheta}-\omega'|} \leq M(B)\int_0^\pi \frac{Bd\vartheta}{\sqrt{B^2+\omega'^2-2B\omega'\cos\vartheta}}\,. \tag{6.13}$$

We continue the estimation by assuming that ω' is positive and taking $\cos\vartheta = 1$. Then it holds that

$$\lim_{B\to\infty}\left|\int_D \frac{N(\Omega)-1}{\Omega-\omega'}d\Omega\right| \le \lim_{B\to\infty} M(B)\int_0^{\pi}\frac{B}{B-\omega'}d\vartheta = \lim_{B\to\infty}\frac{\pi B M(B)}{B-\omega'} = 0\,. \tag{6.14}$$

So far, we have shown that the two first integrals on the right-hand side of (6.9) are zero. Yet we have to calculate the clockwise third integral, which is obtained with the aid of the theorem of residues, which yields that

$$\lim_{\zeta\to 0}\int_{|\Omega-\omega'|=\vartheta}\frac{N(\Omega)-1}{\Omega-\omega'}d\Omega = -\lim_{\zeta\to 0}\mathrm{i}\pi[N(\omega')-1] = -\mathrm{i}\pi[N(\omega')-1]\,. \tag{6.15}$$

Finally we have a Hilbert transform

$$P\int_{-\infty}^{\infty}\frac{N(\Omega)-1}{\omega-\omega'}d\omega = \mathrm{i}\pi[N(\omega')-1]\,. \tag{6.16}$$

Important information needed in the derivation of the K-K relations is the symmetry of the optical constants. This can be observed by consideration of the Fourier transform of the polarization [see (3.7)]

$$P(t) = \epsilon_0\int_0^{\infty}\chi^{(1)}(\omega)E(\omega)e^{-\mathrm{i}\omega t}d\omega\,, \tag{6.17}$$

where the physical reality requires that both the polarization and the electric field are real. This implies that the linear susceptibility has to fulfill the symmetry condition

$$\chi^{(1)}(-\omega) = \left[\chi^{(1)}(\omega)\right]^*\,, \tag{6.18}$$

where "*" denotes the complex conjugate. It follows from (6.4) and (6.18) that the symmetry relations of the real and imaginary parts of the complex refractive index are

$$\begin{aligned} n(-\omega) &= n(\omega)\\ k(-\omega) &= -k(\omega)\,. \end{aligned} \tag{6.19}$$

After some algebra, which is presented in Appendix D, we get finally the celebrated K-K relations

$$\begin{aligned} n(\omega')-1 &= \frac{2}{\pi}P\int_0^{\infty}\frac{\omega k(\omega)}{\omega^2-\omega'^2}d\omega\\ k(\omega') &= \frac{-2\omega'}{\pi}P\int_0^{\infty}\frac{n(\omega)-1}{\omega^2-\omega'^2}d\omega\,. \end{aligned} \tag{6.20}$$

These relations hold for isotropic medium. The static refractive index is obtained from the former relation by setting $\omega' = 0\ \mathrm{s}^{-1}$. In the case of anisotropic medium the relations are more complicated since the tensor nature of the permittivity has to be taken into account [127]. The practical problem with K–K relations is that they require the knowledge of the spectrum in the angular frequency range from zero to infinity. This forms a source of errors in data extrapolations [128,129], which are needed in the calculation of the optical constants beyond the measured range. Most often we wish to calculate the real refractive index change. Then the accuracy of the data inversion can be improved with the aid of multiply-subtractive relations such as those suggested by Goplen et al. [130]. Then the integration interval is finite, but extra information about reference points, i.e. the refractive indices of a liquid at a discrete angular frequency set has to be known. This is usually not a problem since the reference refractive indices can be obtained using ATR.

The success of K–K analysis can be tested using sum rules [131, 132] such as

$$\int_0^\infty [n(\omega) - 1] d\omega = 0 \tag{6.21}$$

and

$$\int_0^\infty \omega k(\omega) d\omega = \frac{\pi}{4}\omega_p^2 \,. \tag{6.22}$$

6.2 Kramers–Kronig Relations for Complex Reflectance

It took considerable long time before K–K relations were formulated for the complex electric field reflectance. The pioneering works are those of Velicky [133] and Smith [134]. Their works and the majority of the studies in this field deal with the reflection of normal incidence light. We follow such a procedure, but first we consider the validity of the assumptions of the K–K relations in the case of oblique incidence, equations (4.1) and (4.2), which can also be the mode of measurement in internal reflection spectroscopy. After showing that the assumptions are valid also for oblique incidence and therefore K–K relations are also valid, we present the derivation of the K–K relations for normal reflectance. We start by dealing with the analyticity of the complex reflectance of (4.1) and (4.2). Since N is analytic it will imply that also that N^2 is an analytic function in the upper-half plane. In the case of liquids, which are insulators, the complex reflectance has no branch points in the upper-half of complex angular frequency plane (which includes the real axis). Then the nominator and denominator of (4.1) and (4.2) are analytic functions. Therefore their quotient is an analytic function except at a finite number of poles, which are located in the lower-half plane (real axis excluded). Such a function is termed in complex analysis as a meromorphic function [125]. The asymptotic fall-off can be observed using the binomial expansions of (4.1) and (4.2). As an example we consider only the case of (4.1). The case of (4.2) can be deduced in a similar manner as shown bellow.

From (4.1) we find that in the case of complex angular frequency variable it holds that

$$r_s(\omega) = \frac{\cos\theta - \sqrt{\epsilon_r(\Omega) - \sin^2\theta}}{\cos\theta + \sqrt{\epsilon_r(\Omega) - \sin^2\theta}} . \tag{6.23}$$

At high angular frequency and using the estimate of (6.7) we get from (6.23)

$$r_s(\Omega) \propto \frac{\cos\theta - \cos\theta - \frac{1}{2}\left(\frac{Re\{\chi^{(1)}(\omega)\}+\mathrm{i}\mathrm{Im}\{\chi^{(1)}(\omega)\}}{\cos\theta}\right)}{\cos\theta + \cos\theta + \frac{1}{2}\left(\frac{Re\{\chi^{(1)}(\omega)\}+\mathrm{i}\mathrm{Im}\{\chi^{(1)}(\omega)\}}{\cos\theta}\right)} \propto -\frac{1}{\Omega^2} . \tag{6.24}$$

As a special case we get the asymptotic behaviour of the normal reflectance from (6.24) by setting simply $\cos\theta = 1$.

The normal reflectance, which can be solved from (6.23), is defined by

$$r(\Omega) = \frac{1 - N(\Omega)}{1 + N(\Omega)} . \tag{6.25}$$

It is certainly analytic function in the upper half plane with appropriate asymptotic properties needed for the existence of the K–K relations. Usually the reflectance of (6.25) is treated in the polar form [see (4.3)]

$$r(\Omega) = |r(\Omega)|e^{\mathrm{i}\varphi(\Omega)} . \tag{6.26}$$

The symmetry relations (6.19) induce the following symmetry relations

$$\begin{aligned} |r(-\omega)| &= |r(\omega)| \\ \varphi(-\omega) &= -\varphi(\omega) . \end{aligned} \tag{6.27}$$

Taking the natural logarithm on both sides of (6.26) makes it possible to resolve the modulus and the phase as follows:

$$\ln r(\Omega) = \ln|r(\Omega)| + \mathrm{i}\varphi(\Omega) . \tag{6.28}$$

For real angular frequency it follows from (6.27) and (6.28) that

$$\ln|r(-\omega)| = \ln|r(\omega)| . \tag{6.29}$$

The use of a logarithmic function seems to be hazardous since apparently it will blow up if the electric field reflectance is equal to zero, which happens for infinite angular frequency. However, in the derivation of the corresponding K–K relations infinity (plus and minus infinity) is approached in a symmetric manner, i.e. taking the Cauchy principal value cancels the blow up of the integral. The usual procedure in context of reflection spectroscopy is the phase retrieval from the intensity reflectance $R = |r|^2$. In other words we wish to resolve the phase for the purpose of calculating the relative complex refractive

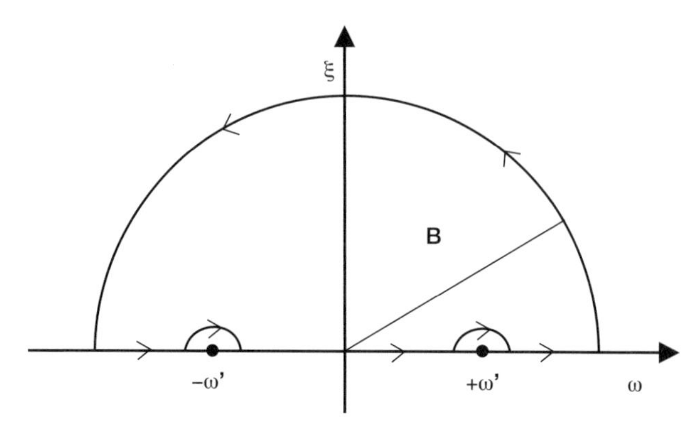

Fig. 6.2. Contours for derivation of Kramers–Kronig relation for phase retrieval from reflectance

index of the liquid. It is rare that we measure the phase and then calculate the intensity reflectance. Therefore, we present a mathematical derivation that results to a K–K relation, which allows the calculation of the phase. The other K–K relation, which gives the reflectance if phase is known, is considered only briefly. We follow the proof given by Smith [134] and define a function

$$f(\Omega) = \frac{\ln r(\Omega)}{\Omega^2 - \omega'^2}, \tag{6.30}$$

which is an analytic function in the upper half plane except at the two poles located on the real axis.

Next we apply to the function f the complex contour integration as shown in Fig. 6.2. Since the mathematical derivation of the Cauchy principal value integral is quite similar to that of Sect. 6.1, the only difference being that one pole more on the real axis has to be handled, we don't write down here the explicit derivation but give the result of the derivation as follows:

$$P\int_{-\infty}^{\infty} \frac{\ln r(\omega)}{\omega^2 - \omega'^2} d\omega = \mathrm{i}\pi \left[\frac{1}{2\omega'} \left(\ln|r(\omega')| - \ln|r(-\omega')| + \mathrm{i}(\varphi(\omega')) - \varphi(-\omega')\right)\right]. \tag{6.31}$$

We next make use of the symmetry relations (6.27) and (6.29) and solve the real and imaginary parts from (6.31). The imaginary part yields the relation

$$P\int_{-\infty}^{\infty} \frac{\varphi(\omega)}{\omega^2 - \omega'^2} d\omega = 0, \tag{6.32}$$

which is only a restatement of the symmetry property of the phase angle. The real part of (6.31) provides the wanted K–K relations, which is

$$\varphi(\omega') = -\frac{2\omega'}{\pi} P\int_{0}^{\infty} \frac{\ln|r(\omega)|}{\omega^2 - \omega'^2} d\omega. \tag{6.33}$$

The other K–K relation can be obtained by similar inspection of another function defined by

$$g(\Omega) = \Omega \ln r(\Omega) \left(\frac{1}{\Omega^2 - \omega'^2} - \frac{1}{\Omega^2 - \omega''^2} \right) . \tag{6.34}$$

Then complex contour integration results to the other K–K relation, which is

$$\ln |r(\omega')| - \ln |r(\omega'')| = \frac{2}{\pi} P \int_0^\infty \omega\varphi(\omega) \left(\frac{1}{\omega^2 - \omega'^2} - \frac{1}{\omega^2 - \omega''^2} \right) d\omega . \tag{6.35}$$

In reflectance data inversion the problem of data extrapolations is also present such as in the case of K–K relations for the complex refractive index. This means that much care has to be taken to fit the wings of the reflectance spectra. To reduce the errors due to data extrapolation Ahrenkiel [135] presented a singly subtractive K–K relation for the purpose of phase calculation. His relation, which provides a better convergence of the integral than the conventional K–K integral, has the expression

$$(\varphi(\omega')/\omega') = (\varphi(\omega_A)/\omega_A) - \frac{2(\omega'^2 - \omega_A^2)}{\pi} P \int_0^\infty \frac{\ln r|(\omega)|}{(\omega^2 - \omega'^2)(\omega^2 - \omega_A^2)} d\omega . \tag{6.36}$$

Here ω_A is an anchor point where the complex refractive index is a priori known and we require that $\omega' > 0$ and $\omega_A > 0$. Multiply subtractive K–K relations [124] yield a better convergence as has been shown by Palmer et al. [136] who studied the distribution of a fixed number of anchor points to obtain most accurate values of optical constants and the optimum number of anchor points. King [137] has presented an efficicient numerical approach to the evaluation of Kramers–Kronig relations. As an example of K–K analysis of reflectance data from liquids we mention the study of Foster [138], who calculated the complex refractive index of nitrobenzene.

Sum rules for the reflectance can be derived [139, 140] but usually they are not of so practical utility as in the case of optical constants.

6.3 Kramers–Kronig Relations in Nonlinear Optics

In the case of nonlinear optics, the interaction of intense light with media is relatively complicated. Nevertheless, the validity of the K–K relations for nonlinear susceptibilities of media has been known for a relatively long time [141–145]. Due to the experimental complexities of nonlinear optics, the validity of the K–K relations was shown as recently as 1992 by Kishida et al [146] for the case of a third-harmonic wave from polysilane film. Bassani and Scandolo [147] were the first researchers who explicitly derived the K–K relations for the optical constants of nonlinear media within the framework of pump and probe configuration.

In the case of so-called nonlinear meromorphic susceptibilities [13, 148] and in femtosecond spectroscopy [149–151], however, it has been shown that K–K relations are not valid. The classical dispersion theory of the linear susceptibility by Lorentz and Drude [9], as well as the quantum mechanical description by Kramers and Heisenberg [9], allows the number of poles of complex linear susceptibility to be countably infinite. Also, the complex linear susceptibility diverges at the poles. We can state that such a complex-valued function behaves like a holomorphic function almost everywhere. In context of linear optical spectroscopy the term "analytic function", which is one of the important properties needed for the existence of the K–K relations, instead of "holomorphic function" has usually been adopted in the literature of this field. Nevertheless, both terms have the same meaning, but in nonlinear optics where two or more complex angular frequency variables may appear the term holomorphic is the appropriate one according to the theory of several complex variables [152]. Now, however, if we consider the linear susceptibility in the whole complex angular frequency plane, it means that we must call the susceptibility a meromorphic function. Fortunately, in the derivation of the K–K relations for normal cases, it is possible to avoid the lower half plane, where the poles of the linear susceptibility are located, by dealing with a function that is holomorphic in the upper half plane. The situation with nonlinear susceptibilities is very similar to the above one. However, nonlinear optical processes that demand the simultaneous existence of poles in both half planes require the generalization of the treatment of the complex nonlinear susceptibility and dispersion relations, because now the nonlinear suscerptibility is a meromorphic function also in the upper half plane.

The equivalence of causality and dispersion relations for linear optical constants was shown by Kronig [153] and later by Toll [123] (see also Nussenzveig [124]. The mathematical essence of the causality is that it provides a necessary and sufficient condition for the existence of K–K relation. There is no doubt about the validity of the K-K relations in the field of linear optics.

In nonlinear optical spectroscopy we usually stimulate nonlinear processes using high-intensity laser beams impinging upon a medium. Then the complex refractive index (see (3.25) can be given for instance in a pump and probe experiment as follows:

$$N(\omega_1, \omega_2; E) = N_L(\omega_1) + N_{\mathrm{NL}}(\omega_1, \omega_2; E) \ , \tag{6.37}$$

where ω_1 is the angular frequency of a weak probe beam, ω_2 is the angular frequency of a pump beam and E is the electric field of the pump beam. Then the K–K relations can be given, using the property of holomorphicity of the complex refractive index in the upper half of complex angular frequency plane and an asymptotic fall off of $N = n + ik$ for high frequency, which is governed by the linear complex refractive index. The K–K relations can be written as follows [147]:

$$n(\omega_1', \omega_2, E) - 1 = \frac{2}{\pi} P \int_0^\infty \frac{\omega_1 k(\omega_1, \omega_2, E)}{\omega_1^2 - \omega_1'^2} d\omega_1 \tag{6.38}$$

$$k(\omega_1', \omega_2, E) = -\frac{2\omega_1}{\pi} P \int_0^\infty \frac{n(\omega_1, \omega_2, E) - 1}{\omega_1^2 - \omega_1'^2} d\omega_1 \,. \tag{6.39}$$

K–K relations similar to the standard forms of (6.38) and (6.39) are valid for most of the optical constants obtained by nonlinear processes. However, if we consider a third-order (or higher) nonlinear single-wave self-action process at frequency ω, then the situation is completely different.

It is obvious that in both (3.26) and (3.28) there appears a minus sign in front of one angular frequency variable. This property has drastic consequences. As we have previously stated in Sect. 3.2 we can describe the angular frequency dependence of the third-order susceptibility with the aid of a function $D(\omega) = (\omega_0^2 - \omega^2 - \mathrm{i}\Gamma\omega)^{-1}$. The third-order nonlinear susceptibility can be expressed using the products of the D functions. Due to the minus sign of the angular frequency, the self-action third-order susceptibility involves a function $D^*(\omega) = (\omega_0^2 - \omega^2 + \mathrm{i}\Gamma\omega)^{-1}$ being the complex conjugate of D. It is evident that D indicates the existence of complex poles of $\chi^{(3)}$ and N in the lower half plane, while D^* indicates the simultaneous existence of poles of $\chi^{(3)}$ and N in the upper-half plane. This means that we have to give up the property of holomorphicity with N, which was a crucial condition for the existence of K–K relations. However, we can now deal with N as a meromorphic function [154], which is holomorphic almost everywhere except on a numerable set of complex poles and zeros. The symmetry properties of the poles was described by Bassani and Lucarini [155]. Now the theorem of residues can be applied to yield dispersion relations, which are non-Kramers–Kronig form [156]

$$\begin{aligned} n(\omega', \omega', -\omega') - 1 = {} & \frac{2}{\pi} P \int_0^\infty \frac{\omega k(\omega, \omega, -\omega)}{\omega^2 - \omega'^2} d\omega \\ & - \mathrm{Im} \left\{ 2\mathrm{i} \sum_{\text{poles}} \mathrm{Res} \left[\frac{N(\Omega, \Omega, -\Omega)}{\Omega - \omega'} \right] \right\} \end{aligned} \tag{6.40}$$

$$\begin{aligned} k(\omega', \omega', -\omega') = {} & \frac{2\omega'}{\pi} P \int_0^\infty \frac{n(\omega, \omega, -\omega) - 1}{\omega^2 - \omega'^2} d\omega \\ & + \mathrm{Re} \left\{ 2\mathrm{i} \sum_{\text{poles}} \mathrm{Res} \left[\frac{N(\Omega, \Omega, -\Omega)}{\Omega - \omega'} \right] \right\} , \end{aligned} \tag{6.41}$$

where the symmetry relations $N_L(-\omega) = N_L^*(\omega)$ and $N_{\mathrm{NL}}(-\omega, -\omega, \omega) = N_{\mathrm{NL}}^*(\omega, \omega, -\omega)$ were used. The residue terms in (6.40) and (6.41), i.e. the series expansions, are calculated for poles located in the first quadrant of the complex angular frequency space. Unfortunately, the residue terms involve complex functions; and such information cannot usually be obtained from measured optical spectra. Furthermore, in order to calculate the residues, we have to know the resonance points of the medium via optical spectrum. This may be rather overwhelming task especially in the case of adjacent overlapping resonance lines in the spectrum.

Causality is always valid. However, it is evident from the above discussion that in a general case, which includes the meromorphic optical constants (see [157] as concerns the causality and dispersion relations of third-order degenerate nonlinear susceptibility), causality is necessary but not a sufficient condition for the existence of K–K relations. An implication of this is that as shown by (6.40) and (6.41) K–K relations are not valid for meromorphic optical constants, and therefore K–K relations are not valid for the corresponding total (nonlinear) reflectance. For total reflectance modified K–K relations analogous to (6.40) and (6.41) can be written [158] but their practical utility in connection with measured spectra is problematic due to complex functions. A situation for the invalidity of K–K relations can be found also in femtosecond spectroscopy in the event of simultaneous incidence of a probe and pump light pulse [146] upon a nonlinear medium. In such a case the third-order nonlinear susceptibility is meromorphic. It is important to emphasize that in cases of femtosecond spectroscopy where pump and probe pulses are exploited the strict causality is broken if the pump arrives before the probe. That is to say, the medium is already responding to a cause (pump) before the probe beam is incident the medium.

The phase retrieval from meromorphic nonlinear susceptibility or reflectance is possible, not by equations analogous to Eqs. (6.40) and (6.41), but by using maxiumum entropy model, which is the subject of the following section.

Finally we mention that sum rules for nonlinear susceptibilities and corresponding complex refractive index can be derived [139, 140, 159].

6.4 Maximum Entropy Method in Phase Retrieval from Reflectance

In optical power spectrum measurements the intensity distribution of light $I(\omega)$ is proportional to the squared modulus $|f(\omega)|^2$ of a complex valued function $f(\omega)$ with a real variable ω. For instance, in reflectance spectroscopy the intensity reflectance $R(\omega) = |r(\omega)|^2$ is measured. Typically, only the modulus of $f(\omega)$ can be measured, although it is necessary to know the complex function $f(\omega) = |f(\omega)| \exp\{i\varphi(\omega)\}$ itself including also the phase $\varphi(\omega)$. A new phase retrieval approach was proposed by Vartiainen et al. [160]. The method is known as the maximum entropy model, MEM. Its basis can be founded in Burg's study [161] of the calculation of power spectra, the data for which consisted of signal measurements taken over a period of time [148, 161]. With its close relation to the concept of maximum entropy, this theory has been also utilized in optical spectroscopy e.g. as a line-narrowing procedure [162, 163].

The applicability of MEM as a phase retrieval procedure has been verified for linear reflectance from solid [164–166] and liquid [167] phases. In the case of the study of [167] the experiments conducted using liquids from the process industry and subsequent comparison with other spectral devices and data analysis methods indicated the correct functioning of MEM analysis with the data obtained using the reflectometer.

In the case of optically nonlinear medium MEM can be applied to phase retrieval from the modulus of the nonlinear susceptibility in the case of anti-Stokes Raman scattering spectrum [168], third harmonic wave from polysilane [169], sum frequency generation spectroscopy [170], meromorphic total susceptibility [171], degenerate nonlinear susceptibility from Maxwell Garnett nanocomposites [172], and reflectivity of nonlinear Bruggeman liquids [40].

The merit of MEM is that it does not require a reflectance determination over the whole electromagnetic spectrum, but only of the region, $\omega_1 \leq \omega \leq \omega_2$, of interest to us. As well as reflectance data, additional information about a given sample is required in order to determine its optical constants. Such information, known as anchor points, commonly comprises the real and/or imaginary parts of the complex refractive index of the sample determined at a frequency within the range $\omega_1 \leq \omega \leq \omega_2$. In practice, MEM phase retrieval procedure includes the experimental reflectance R given by the formula (the somewhat lengthy mathematical derivation can be found in [13])

$$R(\upsilon) = \frac{|\beta|^2}{\left|1 + \sum_{m=1}^{M} a_m \exp(-\mathrm{i}2\pi m\upsilon)\right|^2} , \tag{6.42}$$

where the normalized angular frequency υ is defined by

$$\upsilon = \frac{\omega - \omega_1}{\omega_2 - \omega_1} . \tag{6.43}$$

The unknown MEM coefficients a_m and $|\beta|$ can be obtained from a set of linear Yule–Walker equations [173]

$$\sum_{m=0}^{M} a_m C(n-m) = \begin{cases} |\beta|^2 & n = 0 \\ 0 & n = 1, \ldots, M \end{cases} , \tag{6.44}$$

where the auto-correlation $C(t)$ is computed by a Fourier transform of $R(\upsilon)$ as

$$C(t) = \int_0^1 R(\upsilon) \exp(\mathrm{i}2\pi t\upsilon) d\upsilon . \tag{6.45}$$

Phase retrieval can now be based on MEM for complex reflectivity

$$r(\upsilon) = \frac{|\beta| \exp[-\mathrm{i}\phi(\upsilon)]}{1 + \sum_{m=1}^{M} a_m \exp(-\mathrm{i}2\pi m\upsilon)} . \tag{6.46}$$

In (6.46) the error phase $\phi(v)$ is the only quantity that cannot be obtained by measurement of $R(v)$. The idea of using MEM in phase retrieval is that the problem of finding the true phase $\varphi(v)$ is reduced to a problem of finding the error phase $\phi(v)$ (typically, ϕ is a much simpler function than φ). The additional information on $r(v)$ determined at $L+1$ discrete frequencies v_l is used to find an estimate for $\phi(v)$

$$\phi = \sum_{l=0}^{L} B_l v^l \ , \tag{6.47}$$

where B_l is obtainable from a Vandermonde system

$$\begin{pmatrix} 1 & v_0 & \cdots & v_0^L \\ 1 & v_1 & \cdots & v_1^L \\ \vdots & \vdots & & \vdots \\ 1 & v_L & \cdots & v_L^L \end{pmatrix} \begin{pmatrix} B_0 \\ B_1 \\ \vdots \\ B_L \end{pmatrix} = \begin{pmatrix} \phi(v_0) \\ \phi(v_1) \\ \vdots \\ \phi(v_L) \end{pmatrix} . \tag{6.48}$$

The error phase is usually a slowly varying function and in favourable cases only one or two anchor points are needed, i.e. the optimum degree of the polynomial is low. Brun et al. [174] have presented an optimization method in order to improve the error phase smoothening of a reflection spectrum. Vartiainen et al. [175] derived sum rules in testing non-linear susceptibility obtained using the maximum entopy model.

As an example we show here a simulation based on the application of MEM into phase retrieval from wavelength-dependent surface plasmon resonance dip. The application of the above numerical method to the calculation of the complex permittivity (4.53) is demonstrated in Fig. 6.3. In the simulation optically dense liquid, obtained by diluting red food colour in water, was used as a sample. The SPR dip of this liquid is shown in Fig. 6.3a, and the ME phase in Fig. 6.3b together with the location of four anchor points. In Figs. 6.3c and d are shown the MEM estimates of the real refractive index and the corresponding extinction coefficient obtained with the aid of (3.11). The real refractive index is usually obtained with great accuracy. However, the extinction coefficient is susceptible to a larger error because equation (4.53) tends to emphasize small errors of the reflectance R_p.

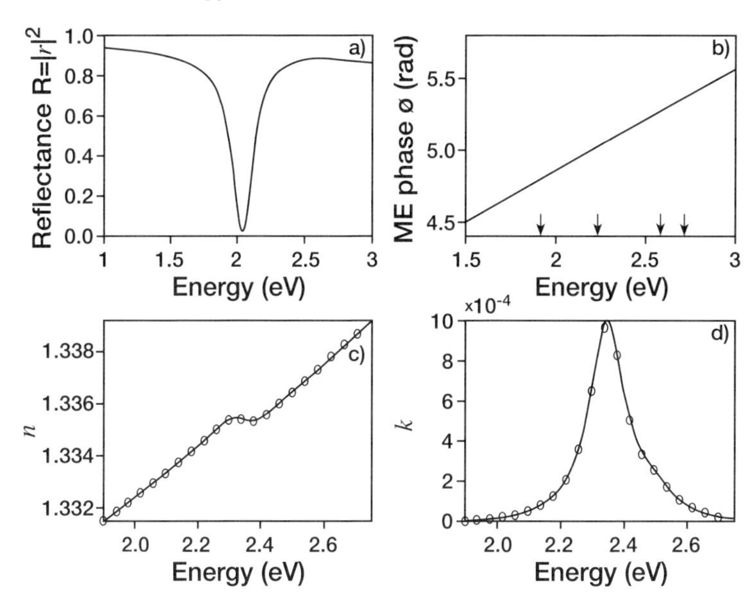

Fig. 6.3. (**a**) Reflectance of red color liquid, (**b**) ME phase, and (**c**) real and (**d**) imaginary parts of the complex reflectivity (solid lines) and corresponding MEM estimates (circles). The arrows in (**b**) indicate the energies at which the phase is a priori known. The angle of light incidence is 70°

Part II

Practical Reflectometry

7 Introduction

As the title of this book suggests the wavelength range of the electromagnetic radiation we have employed is restricted to the ultraviolet/visible (UV/Vis) range. For natural reasons there is a long tradition in science of exploiting this range of wavelengths, particularly those of visible light. Indeed, the human eye was for a long time the principle instrument of observation in research and experimental work until the appearance of manmade detectors such as thermometers and photographic plates. Such detectors made it possible to research and exploit areas of the electromagnetic spectrum extending far beyond that perceivable by the naked eye. At the same time the detected signal could be measured against some suitable absolute scale. The foundation of quantitative measurement in UV/Vis spectroscopy was laid in the mid-1800s with the appearance of the Bourguer-Lambert-Beer law. This law demonstrates the relationship between the ability of a sample to absorb electromagnetic radiation and the composition of the sample.

The UV/Vis range is still important and frequently used in the natural sciences. Optical information in this range is created when the outermost electrons of atoms and molecules move from one energy level to another as the result of interaction between electromagnetic radiation and the material in question. Although the spectra of liquids and solid matter are not as detailed as those of gases or IR spectra, the analysis of UV/Vis spectra nevertheless provides information on the characteristics of the material under examination as well as the relative proportions of its constituent parts. Numerous applications may be found especially within chemistry – the quantitative analysis of one or more elements and of organic compounds, water analysis, enzyme analysis and kinetics, investigation of chemical equilibria and the analysis of chemical reactions, etc. [176]. Optical techniques exploited in industry has been enlightened e.g. by Cielo [177].

With regard to present-day instrument technology and therefore also to the measurement procedures nowadays employed the UV/Vis range is generally understood as extending from 200 to 1000 nm and therefore partly includes the near infrared area (NIR). Indeed, many commercial spectrophotometers operate specifically in this range. The limits of the range are naturally determined by the detectors and light sources employed as well as the materials brought into contact with the radiation. Commonly employed con-

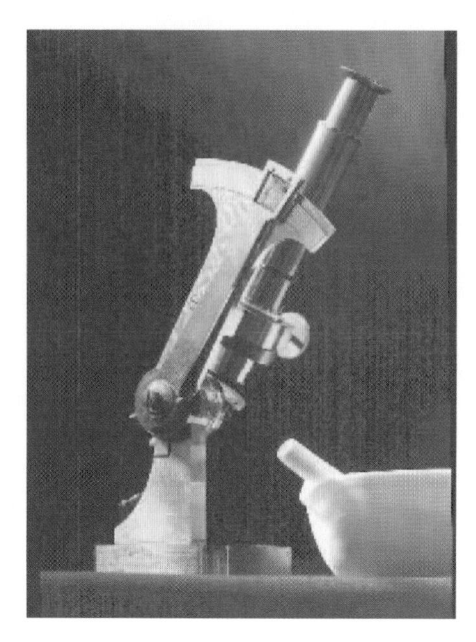

Fig. 7.1. A photograph of one of the very first refractometers from the 1800s. This particular apparatus was capable of measuring the refractive index of glass material. (Photo courtesy of Karl Zeiss Jena GmbH)

temporary semiconductor-based detectors, i.e. photodiodes, typically operate in this range. Atmospheric nitrogen also leaves its own mark in the region of the spectrum employed – electromagnetic radiation below 190 nm is efficiently absorbed by nitrogen. Of course, measurements of these wavelengths may be performed in a vacuum.

The subject of this book deals with the theory and the measurement of reflectance. In brief it is the study of radiation reflected from a sample of matter (usually from its surface) in order to provide information on the chemical and physical characteristics of the material and of changes taking place therein. Such study often makes use of particular regions of the spectrum.

Science has a long tradition of using the principle of reflectance. Figure 7.1, for example, reproduces a measurement device from the end of the 1800s. The photograph is of a refractometer developed by Ernst Abbe. This device was used to determine the refractive index of solids.

The reflection and refraction-based principle of the refractometer, i.e. determination of the so-called critical angle, is still applied in corresponding contemporary devices such as the refractometer developed for use in industrial process conditions and presented in Fig. 7.2.

The reflection method is still in use, especially in the qualitative and quantitative analysis of solid matter. Subjects of study include crystalline powder, pigments and colouring agents. The method is also used in the study and

Fig. 7.2. A modern on-line refractometer designed for process measurements. The system consists of a sensor and an indicating transmitter. (Photo courtesy of K–Patents)

determination of molecular structure, adsorption, surface catalysis, reaction kinetics, photochemistry and optical constants [51]. The main interest of the present authors is the exploitation and development of reflection measurements in order to gain information on the characteristics of various kinds of liquids, in particular industrial process liquids. Such measurements would enable the determination of, for example, the optical constants of solvents, i.e. the refractive index n and extinction coefficient k. These may be further combined with more directly observable and accessible quantities such as solids content, concentration, density and even temperature.

The second part of this book examines the practical side of reflectance measurement. Subjects dealt with include measurement procedures and related issues such as calibration and noise control. We also delve inside the workings of a spectrophotometer in order to ascertain its most important parts and components and how these function. These matters are presented with an emphasis on optics and opto-electronics while less attention is paid towards the actual electrical functioning of the devices, associated electronics and mechanical solutions. Additionally, we give the results used to test out the applicability of the theories presented at the beginning of the book. All the measurements presented have been obtained using a reflectometer designed for research use and developed and built by the authors.

8 Definitions of Optical Instrumentation and Measurement

In this chapter we turn our attention to the definitions, terms, quantities and units most commonly employed in optical instrumentation and measurements. For example, in the planning and construction of experimental arrangements and, in particular, in the application of light sources and detectors there is a need for a basic understanding of *radiometry* and *photometry*. We will also define some basic concepts such as *absorptance*, *transmittance* and *reflectance*. There is still a plethora of terms in use in this field of science. Indeed, terminology developed for use in one discipline is sometimes carelessly applied in another. Our treatment of this subject follows the thorough expositions by Zalewski [178] and Palmer [179].

8.1 Radiometry

The measurement of optical radiation requires an understanding of the physical principles of radiometry as well as of the measurement equipment in use in this field. Radiometry examines the energy of electromagnetic fields and how this energy is transferred from the source, generally through a non-dissipating medium, to the detector. It should be noted that during such measurement no account is taken of possible effects of the radiation on the sample or detector itself. Furthermore, the transfer of energy is often assumed to obey only the laws of geometrical optics while optical phenomena such as interference, diffraction or the degree of coherence are disregarded. The performance of radiometric measurements across the entire spectrum requires the use of a variety of techniques and materials. On this basis radiometry may be divided into a number of branches according to the wavelength range to be examined. Thus we have, for example, microwave radiometry and visible light radiometry.

The electromagnetic radiation may be categorized as shown in Fig. 8.1. The figure demonstrates that the familiar visible light covers only a narrow wavelength band of the whole electromagnetic radiation. The wavelengths defining the categories are not strict, indeed the regions are in some extend overlapped. Furthermore, the regions are often subdivided. For example, the infrared range consists of near, intermediate, far and extreme infrared regions.

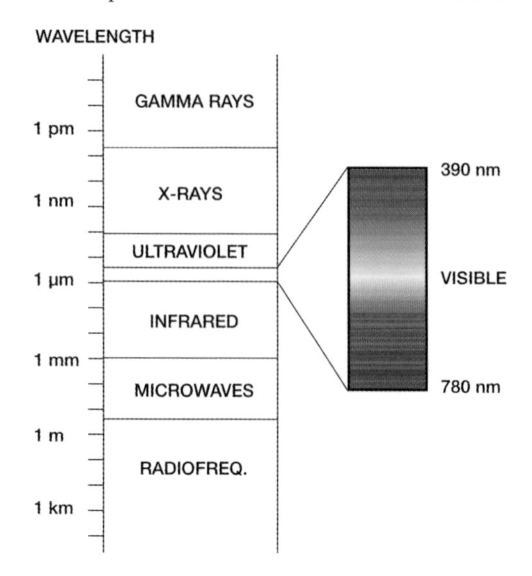

Fig. 8.1. The electromagnetic-photon spectrum

The UV and visible ranges cover wavelengths roughly from 10 to 780 nm and the limit between these ranges lies at 390 nm. UV-visible radiation originates mainly due a rearrangement of the outer and the inner electrons in atoms and molecules. This radiation is observed by electrical detectors such as photomultiplier tubes and the visible light, naturally, by the eye. Photons energies varies from 1.6 to 3.2 eV or alternatively $2.5 \cdot 10^{-19}$ to $5.1 \cdot 10^{-19}$ J at visible range. Although a small energy of an individual photon as few as ten photons arriving the eye may cause a visual sensation.

The basic quantities used in radiometry are *radiant energy* Q and *radiant power* Φ. Radiant energy is the energy which is emitted, transferred or received in electromagnetic form. Correspondingly, radiant power (or radiant flux) is the power which is emitted, transferred or received in electromagnetic form. The relationship between these two is naturally given by

$$\Phi = \frac{dQ}{dt} \,. \tag{8.1}$$

If our source emits radiation continuously and evenly, in other words, if the output of the source is independent of time, so the measurement of the radiation will relate to the measurement of power Φ. In this case the unit employed is the watt (W). Correspondingly, the measurement of flashing radiation or individual pulses relates to the measurement of energy Q, in which case the unit employed is the joule (J). The choice of these units is based largely on tradition and earlier level of technology – radiometric measurements were often employed during measurements of thermal energy and power. However, in many applications the measurement of the numbers of photons provides

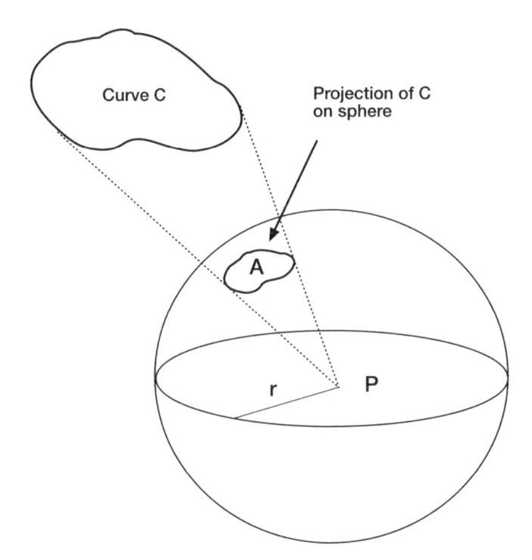

Fig. 8.2. An arbitrary curve C and a point P define the solid angle Ω. The magnitude of Ω corresponds the area A, which is obtained by the radial projection of the curve C to the surface of a sphere of unit radius. The unit of the solid angle is the steradian (sr)

physically more sensible results as a consequence of the quantum character of various (e.g. photochemical and photophysical) phenomena.

The *solid angle* Ω is defined by the closed curve C and point P in space as shown in Fig. 8.2. The magnitude of the solid angle is described by the area A, which is obtained by the radial projection of the curve C to the surface of a sphere of unit radius and with central point P [180]. In other words, if dA is an infinitesimal area on the surface of a sphere of radius r, so

$$d\Omega = \frac{dA}{r^2} . \tag{8.2}$$

Although according to this equation Ω is a dimensionless quantity, it is generally assigned a unit, namely the steradian (sr).

Irradiance E and *exitance* M both describe the density of radiant power on an arbitrary surface. We use irradiance E if the surface receives radiation (such a surface may generally be called a detector). Correspondingly, if radiation is emitted from the surface (thus acting as a source) we use the term exitance M. These are defined as follows

$$E = \frac{d\Phi}{\cos\theta_d dA_d} \tag{8.3}$$

and

$$M = \frac{d\Phi}{\cos\theta_s dA_s} . \tag{8.4}$$

In these formulae θ is the angle between the normal of the surface and the direction of the radiation, and dA is an infinitesimal area on the surface.

The sub-indexes d and s refer to the detector and source, respectively. Both quantities are measured in units of watts per square metre ($\mathrm{W/m^2}$).

Radiant intensity I (or simply intensity) describes the radiant power of a source in any given direction with respect to the solid angle. Thus I is given by

$$I = \frac{d\Phi}{d\Omega} \,. \tag{8.5}$$

The unit employed is the watt per steradian (W/sr). Strictly speaking intensity is defined only for a point source. As a consequence all intensity measurements are to a greater or lesser degree approximations.

In physical optics the letter I is also understood as the time-averaged value of the magnitude of the *Poynting vector* S according to the equation [181]

$$I \equiv \langle S \rangle_T = \frac{c\epsilon_0}{2} E_0^2 \,. \tag{8.6}$$

Here c is the speed of light in a vacuum, ϵ_0 the permittivity of the vacuum, E_0 the amplitude of the electric field of the electromagnetic wave and T refers to the intergation time. Thus defined I corresponds with radiometric irradiance and is proportional to the square of the amplitude of the electric field.

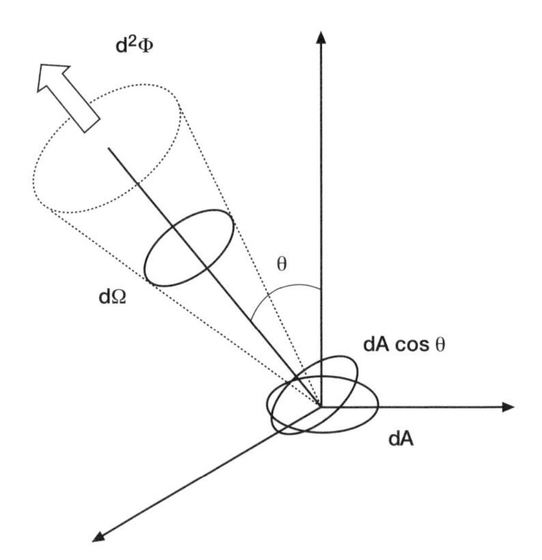

Fig. 8.3. Geometry for describing the concept of radiance. Radiance in terms of visual response may be characterised by the brightness [182]

In studies in the field of geometrical optics and especially in ray tracing use may be made of the *Optical invariant*[1] (or *Lagrange invariant*). It may be demonstrated that this remains unaltered throughout the entire optical system under examination [183]. Correspondingly, in radiometry the term *radiance* L retains its value on moving from one point to another within a lossless medium. Radiance is measured in units of (W/sr/m^2) and is calculated using the equation (see Fig. 8.3)

$$L = \frac{d^2\Phi}{\cos\theta_s dA_s d\Omega} . \tag{8.7}$$

In radiometry a source whose radiance L_0 is independent of the viewing angle θ is called *Lambertian source.* Several non-coherent real sources such as solids at high temperature or white paper approximate this criteria of constant radiance. Therefore the concept of Lambertian source is important. For a planar Lambertian source it can be shown that the intensity is

$$I = A_s L_0 \cos\theta . \tag{8.8}$$

Thus radiation follows the *cosine law* i.e. $I = I_0 \cos\theta$. Often radiometric measurements deal with sources (and detectors) of circular shape. If radius of such source is r the irradiance E at the distance s is

$$E = \begin{cases} \pi L_0 & s \ll r \\ \frac{\pi r^2 L_0}{s^2} & s \gg r . \end{cases} \tag{8.9}$$

In the above study spectral characteristics of the radiation was not dealt with. *Spectral radiant power* distribution is represented by Φ_λ and the unit most commonly used is (W/nm). Correspondingly, radiant power may be studied with relation to frequency f. In this case the unit employed is (W/Hz). Since the wavelength and frequency of an electromagnetic wave are mutually dependent it may be shown that $\lambda\Phi_\lambda = f\Phi_f$. The radiant power Φ of the entire spectrum may be obtained by summing the radiant powers Φ_λ of all wavelength intervals

$$\Phi = \int_0^\infty \Phi_\lambda d\lambda . \tag{8.10}$$

The spectral quantities associated with other radiometric quantities such as radiant energy and radiance are determined correspondingly.

---

[1] The invariant $\mathcal{I}$ of an optical system may be defined by two particular rays namely the axial and the oblique rays and the height of the object in question. Thus $\mathcal{I} = y_o n u_a - y_a n u_o$. Here n is the refractive index of the medium, y is the object height and u is the angle between the optical axis of the system and the oblique or the axial ray. Subscripts "o" and "a" denote to oblique and axial rays, respectively.

8.2 Actionometry

The production of a signal in radiometry is often based on the thermal effect of radiation. If we wish to emphasise the particle nature of electromagnetic radiation it would be wise to use quantities measured in units of photons or photons per second. In addition, if we are not required to take account of possible photophysical, photochemical or photobiological processes caused by radiation we may use the term actionometry. The basic quantity of actionometry is *photon flux* $\mathcal{N}$. The photon flux for monochromatic radiation $\mathcal{N}_\lambda$ is given by

$$\mathcal{N}_\lambda = \beta_1 n \lambda \Phi_\lambda \,, \tag{8.11}$$

where β_1 is a constant ($\beta_1 = 5.0341 \cdot 10^{15}$), n is the refractive index of the medium and Φ_λ is the radiant power at wavelength λ (given in nanometres). Correspondingly, if the sample is bombarded with radiation for a period of time t the *photon dose* U may be calculated using the formula

$$U = \beta_2 n \lambda \Phi_\lambda t \,, \tag{8.12}$$

where β_2 is a constant ($\beta_2 = 8.3593 \cdot 10^{-9}$). Another practical unit used in actionometry and particularly in photochemistry is known as the *Einstein*, which indicates the amount of energy in one mole ($6.022 \cdot 10^{23}$) of photons.

8.3 Photometry

Photometry and radiometry are closely related to one another and are distinguished solely by the spectral response of the detector employed. In the case of photometry the detector used operates in the range of visible light. Photometry involves measuring the ability of electromagnetic radiation to give rise to a physically observable visual sensation. Since visual sensation differs from person to person we require a previously arranged and defined average model for visual ability. Such a definition has been performed by the CIE.[2]

Before the development of present-day detectors the naked eye was of considerable importance in physical measurements – even though the eye was unable to perform absolute measurements it was still being used as a detector in precise photometric measurements in the second half of the twentieth century. The importance of photometry is highlighted by the fact that one of the basic units of the SI system (Système International), the *candela* (*cd*), relates directly to human physiology.

[2] CIE, the *Commission Internationale de l'Eclairage*, is an international body founded in 1913 which regulates the terminology and basic concepts employed in radiometry and photometry as well as developing standards and measurement procedures in these fields.

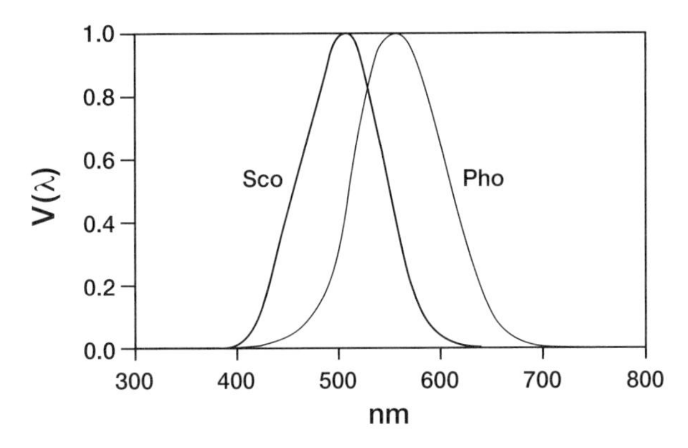

Fig. 8.4. The response curves $V(\lambda)$ for photopic and scotopic vision. The curves show that the maximum sensitivities of the cones and rods in the eye are located at different wavelengths

Photometry is based on a knowledge of the functioning of the human eye. This is no surprise as the photometric detection system has to be comparable with the average range of human visual perception i.e. the CIE standard observer. In practice photometry is founded on radiometry – radiometric measurements are modified into photometric quantities using the so-called *spectral efficiency function.*

Of the two types of cells in the eye which are sensitive to light those known as cones are of greater importance when the intensity of light is sufficient (i.e. when the radiant power is high). In such cases we speak of *photopic vision.* As radiant power decreases the cells known as rods become activated and work together with the cones (*mesotopic vision*). In conditions of considerable darkness only the rods are stimulated and their functioning is described as *scotopic vision.* The response curves for photopic and scotopic vision as determined by the CIE are presented in Fig. 8.4.

The quantity of response of radiometric radiant power in photometry is known as *luminous flux* Φ_v. Its unit is the *lumen* (lm). It describes spectral radiant power in terms of visual response. In photopic vision the luminous flux is given by

$$\Phi_v = K_m \int \Phi_\lambda V(\lambda) d\lambda \, , \tag{8.13}$$

where K_m is the *luminous efficacy* for phototopic vision ($\cong 683\,\mathrm{lm/W}$) and $V(\lambda)$ is the spectral luminous efficiency function. Similar equation can be derived for scotopic vision. Then the luminous efficacy $K'_m \cong 1700\,\mathrm{lm/W}$.

Illuminance E_v is the photometric equivalent to irradiance E in radiometry. It is measured in units of ($\mathrm{lm/m^2}$) and is determined using the formula

$$E_v = \frac{d\Phi_v}{\cos\theta_d dA_d} \, . \tag{8.14}$$

Radiant intensity I finds its counterpart in *luminous intensity* I_υ. The unit employed is the *candela* (cd) or, alternatively, lumen per steradian (lm/sr).

$$I_\upsilon = \frac{d\Phi_\upsilon}{d\Omega} . \tag{8.15}$$

Likewise, radiometry's radiance L is replaced in photometry by *luminance* L_υ which is measured using the unit of candela per square metre (cd/m^2)

$$L_\upsilon = \frac{d^2\Phi_\upsilon}{\cos\theta_s dA_s d\Omega} . \tag{8.16}$$

Table 8.1 provides a summary of the most important quantities used in radiometry, photometry and actionometry. Further reading may be found in references [184–186].

Table 8.1. Basic quantities employed in radio-, photo- and actinometry [178]

	Radiometric	Fotometric	Actinometric
Base quantity	Radiant power or radiant flux (W)	Luminous flux (lm)	Photon flux (photons/s)
Surface density	Irradiance (W/m^2)	Illuminance (lm/m^2)	Photon flux irradiance
Solid angle density	Radiant intensity (W/sr)	Luminous intensity (cd) or (lm/sr)	Photon flux intensity
Solid angle and surface density	Radiance (W/sr/m^2)	Luminance (cd/m^2) or (lm/sr/m^2)	Photon flux radiance

8.4 Spectrophotometry

Spectrophotometry deals with the effect of the medium and matter on the transfer of electromagnetic radiation. The basic processes involved here are *reflection*, *absorption* and *transmission* of radiation. Observation of these phenomena offers the possibility not only of examining the transfer of energy but also the analysis of matter.

When radiation strikes the sample shown in Fig. 8.5, for example, all three phenomena – reflection, absorption and transmission – occur. Reflection comprises of one of two extreme types: *specular reflection* and *diffuse reflection* [51]. In the latter case light is scattered so profoundly that when reflected from the surface of a flat object it fills an entire hemisphere. In reality reflection always comprises some combination of these two extreme types. The drawing in Fig. 8.5 shows the theoretical extremes of reflection and transmission.

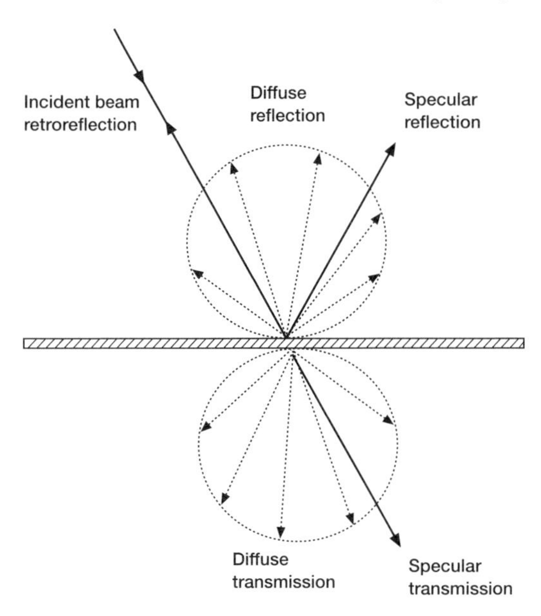

Fig. 8.5. When a light beam encounters a material, three basic phenomena may be observed, i.e. reflection, transmission and absorption. The extreme cases of reflection are specular and diffuse reflections. The same limiting cases may be applied to transmission

8.4.1 Reflectance

As may be seen in Fig. 8.5 a portion of the radiation returns to the space from which it was emitted. This phenomenon is known as reflection. Reflection may be specular, diffuse or some combination of the two. *Reflectance* ρ may be defined over a wide wavelength area as the ratio of incoming Φ_i and reflected Φ_r radiant power

$$\rho = \frac{\Phi_r}{\Phi_i} \,. \tag{8.17}$$

Correspondingly, reflectance[3] may be defined for a given wavelength. In this case it is denoted by the term $\rho(\lambda)$. In publications from the field of optics reflectance is often denoted using the letter R (a practice the present authors have adopted elsewhere in this book). It should be noted that R is also reserved for expressing to what extent the radiant flux reflected from the

[3] There has been discussion concerning when the endings of terms -ivity and -ance (e.g. transmittivity and transmittance) should be employed. Nowadays the following practice is recommended. If we are referring to a pure material we use the former-mentioned ending. Hence, the reflectivity of pure silver. The latter-mentioned ending is used when speaking generally characteristics of a sample, for example, the reflectance of a sample of aluminium coated with oxide.

sample diverges from reflection from an ideal diffusing surface. In this case we speak of the *reflectance factor.*

8.4.2 Transmittance

The term transmission refers to an event in which the radiant flux incident to the surface of a sample departs from a different surface (usually from the opposite side of the sample). In this case *transmittance* τ is determined by

$$\tau = \frac{\int_0^\infty \tau(\lambda)\Phi_{\lambda i} d\lambda}{\int_0^\infty \Phi_{\lambda i} d\lambda} . \tag{8.18}$$

Here $\Phi_{\lambda i}$ is the incident radiant flux and $\tau(\lambda)$ is the *spectral transmittance* of the sample which is given by the formula

$$\tau(\lambda) = \frac{\Phi_{\lambda t}}{\Phi_{\lambda i}} . \tag{8.19}$$

Correspondingly, $\Phi_{\lambda t}$ is the radiant flux which is transmitted through the sample. Transmittance is also commonly represented by the letter T.

Materials may be classified according to their reflectance/transmittance characteristics. Materials yielding values of $\tau = 0$ are said to be exclusively reflecting. These include objects such as mirrors as well as enamelled and varnished surfaces. These latter-mentioned surfaces produce high degrees diffusion while reflection in mirrors is specular. The value of $\tau = 0.35$ may be taken as the boundary between two further classifications. Materials with lower transmittance are said to be weakly transmitting and strongly reflecting. Examples of such materials include paper, textiles and colour filters. τ values in excess of 0.35 indicate strongly transmitting materials such as window glass, opal glass and plastic sheet. We note that this latter group includes both highly and weakly diffusing materials.

8.4.3 Absorptance

The interaction between electromagnetic radiation and matter may cause a part of the radiation to change into a different form of energy, usually heat. This third major process is known as absorption. Its associated quantity, *absorptance* may be determined using the equation $\alpha = \Phi_a/\Phi_i$ where Φ_a represents the amount of incoming radiation absorbed by the material. Absorptance may also be determined using the equation

$$\alpha = \frac{\int_0^\infty \alpha(\lambda)\Phi_{\lambda i} d\lambda}{\int_0^\infty \Phi_{\lambda i} d\lambda} , \tag{8.20}$$

where $\alpha(\lambda)$ is the *spectral absorptance* and $\alpha(\lambda) = \Phi_{\lambda a}/\Phi_{\lambda i}$.

The energy transformation process is also described by the *absorption coefficient* α'. This may be calculated using the formula for transmittance

$$\tau_{int} = \exp(-\alpha' d) \ , \tag{8.21}$$

where τ_{int} is the internal transmittance of an object of material thickness d. The unit by which the absorption coefficient is measured is the inverse of the distance travelled. One commonly used unit is (cm^{-1}).

The use of transmittance with highly absorbing samples is not particularly revealing. *Absorbance* A describes the logarithmic damping of light in a sample and may be defined in terms of transmittance using [176, 187]

$$A = \log_{10}(1/T) \ . \tag{8.22}$$

An alternative name for absorbance is *optical density* D.

It may be shown that reflectance, transmittance and absorptance occupy a mutual relationship given by

$$\rho + \tau + \alpha = 1 \tag{8.23}$$

while the spectral quantities obey

$$\rho(\lambda) + \tau(\lambda) + \alpha(\lambda) = 1 \ . \tag{8.24}$$

These equations are a consequence of the *law of the conservation of energy.* In the latter formula (8.24) it is supposed that the processes involved to not include any non-linear phenomena such as the Raman effect.

8.5 Reflectance Measurement Geometries

Radiation may be brought to the sample using various means, for example, as unidirectional radiation or in a conical shaped solid angle. At the same time reflected radiation may be collected and observed in a variety of ways. From this it follows that there are numerous geometries relating to reflectance and its measurement. Two particular geometries, directional-hemispherical reflectance and conical-directional reflectance, are presented in Figs. 8.6 and 8.7. In the diagram 8.6 a unidirectional beam of light arrives at the sample. The direction of the beam may be determined using the polar angle θ and azimuth ϕ. Reflectance is observed throughout the solid angle 2π, in other words, over the entire hemisphere. In the diagram 8.7 the incoming radiation is focused into a cone shape. Here light reflected is examined in one specific direction at a time.

A fundamental descriptor of reflectance geometry is the *bidirectional reflectance distribution function (BRDF).* This is assigned the symbol f_r. BRDF is given by the equation

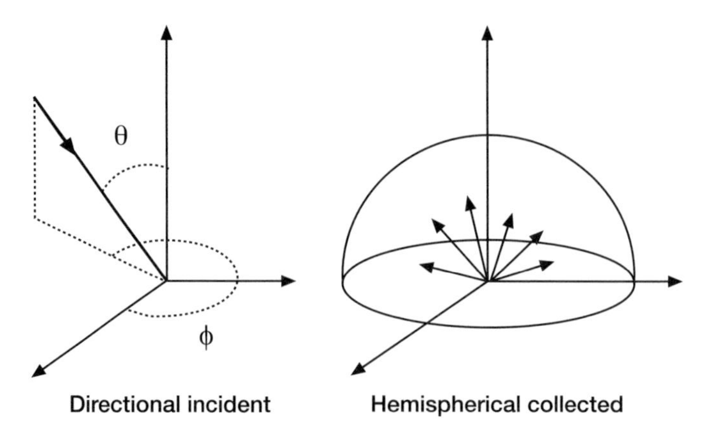

Fig. 8.6. An example of reflectance geometry, i.e. directional-hemispherical reflectance. Reflectance geometries may be described by the bidirectional reflectance distribution function (BRDF)

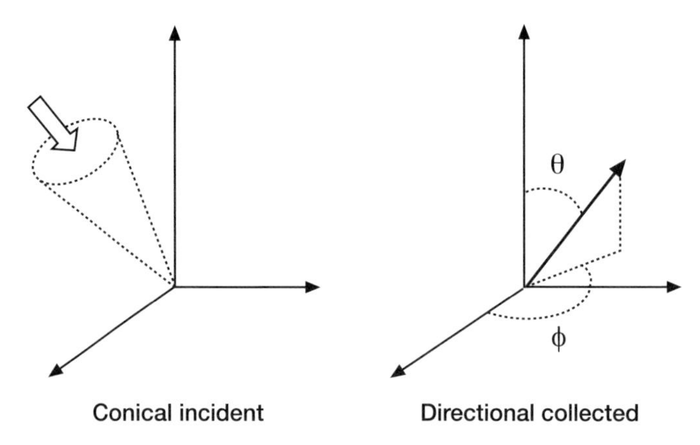

Fig. 8.7. Conical-directional reflectance geometry

$$f_r(\theta_i, \phi_i, \theta_r, \phi_r) = \frac{dL_r(\theta_i, \phi_i; \theta_r, \phi_r; E_i)}{dE_i(\theta_i, \phi_i)} , \tag{8.25}$$

where dL_r and dE_i are the differential elements of the reflected radiance and radiant incidence. Both are defined for a given direction (see e.g. Fig. 8.7) – the polar angle θ is determined in relation to the normal to the surface and the azimuth ϕ in relation to some convenient reference point on the surface. The unit employed with the BRDF is the (sr^{-1}).

In the example of directional-hemispherical reflectance given in Fig. 8.6 we obtain the following equation for reflectance

$$\rho(\theta_i, \phi_i, 2\pi) = \int_{2\pi} f_r(\theta_i, \phi_i, \theta_r, \phi_r) d\Omega_r . \tag{8.26}$$

Examples of other reflection geometries along with additional reading material may be found in references [179, 188, 189].

9 Exploring the Insides of a Spectrophotometer

The functioning of optical measurement devices is based on the observation of changes and effects produced by the interaction of electromagnetic radiation and matter. From a technical viewpoint this requires the generation of electromagnetic radiation, the modification of its characteristics and the control of its propagation through a given space. Similarly, the "fingerprints" which the material leaves on the radiation need to be producible in an easily understood form. In this often complicated process use is made of optics, optoelectronics, electronics and mechanics. In this chapter we shall examine the nature and application of the main components used in spectroscopic instruments. These include light sources, components and devices for selecting the desired wavelength, polarizers and detectors.

Before considering these components in detail we will briefly examine an example of a spectrophotometer, the optical principle of which is presented in Fig. 9.1. Although this drawing is not a exact copy of any existing commercial instrument it nevertheless contains all the basic functions and components found in a modern *double-beam spectrophotometer*. It has two light sources, i.e. a deuterium lamp and a halogen lamp for performing measurements across the UV/visible range. The light source is selected using a rotating mirror. The useful absorbance range has been made as wide as possible by effectively eliminating stray light using two monochromators (here the monochromator is made up of two slits and a grating). Following these the light path is split into a reference beam and a sample beam using a perforated rotating mirror. The beams are directed into a sample chamber and thereafter into a photomultiplier. When a part of the radiation is absorbed by the sample an alternating signal is recorded by the detector. The magnitude of this variation is then converted into the transmittance value. By scanning the wavelength the spectra of the sample are provided.

The components and light paths of the instrument are shielded from external light and dust. Spectrophotometric measurements may also be performed using only one monochromator or without the reference light path. In the latter case the instrument is known as a *single-beam spectrophotometer*.

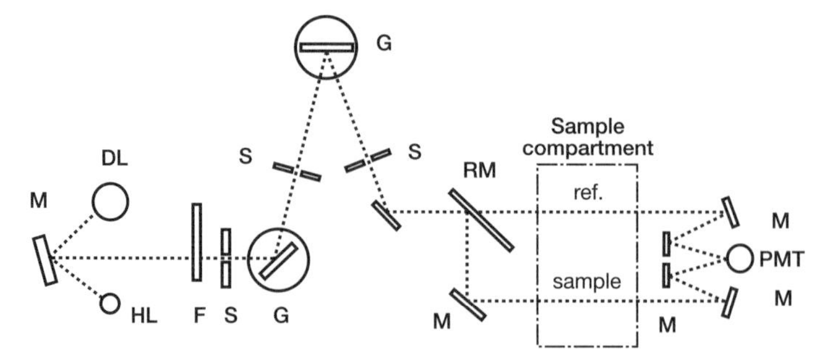

Fig. 9.1. The layout of an imaginary spectrophotometer. The optical path is indicated by a dotted line. Components are denoted as follows: HL = halogen lamp, DL = deuterium lamp, M = mirror, F = filter wheel, S = slit, G = grating, RM = rotating mirror, PMT = detector (photo multiplier tube)

9.1 Common UV/Visible Range Light Sources

Electromagnetic radiation in the UV/visible range is generated by suitable light sources. A good light source should produce sufficient energy or power, the radiation it generates should be stable or regularly altered over time and its spectral distribution should be suitable for the application in question. In addition, the light source should be hard-wearing, long-lived and easy to replace. Lasers produce considerable power over a narrow waveband and have for this as well as other reasons become an increasingly popular light source in spectroscopy. However, a detailed study of the functioning and characteristics of lasers is beyond the scope of the present work.

Before the development of the modern-day range of electric powered light sources researchers exploited our nearest star, the sun. For example, in studies performed at the end of the 1800s A.G. Bell made use of sunlight in his pioneering experimental observations of the so-called *photoacoustic effect* [190]. At the same time he created the foundation for *photoacoustic spectroscopy*. Although a number of problems are associated with the use of the sun as a light source it is, nevertheless, a good example of a light and energy source which produces radiation over a wide spectral range.

Although visible light and electromagnetic radiation in general may be produced by numerous techniques the generation of light is primarily based on the relaxation process of the excited states of atoms and molecules. Figure 9.2 shows schematically the main transitions which may occur in atom (without taking the vibrational states into account). Those relaxation processes which generate radiation are indicated by dotted line in the figure. S_0 represents the ground energy level for an electron. All other states are excited, either singlet (S) or triplet (T), states. Non-radiative transitions (solid lines) include internal coversion (XI, XII and XIII) and intersystem crossing (XIV and XVII). The transitions V and VI refer to fluorescence and phosphorescence. A thorough treatment of the issue may be found e.g. in [191]. It should be

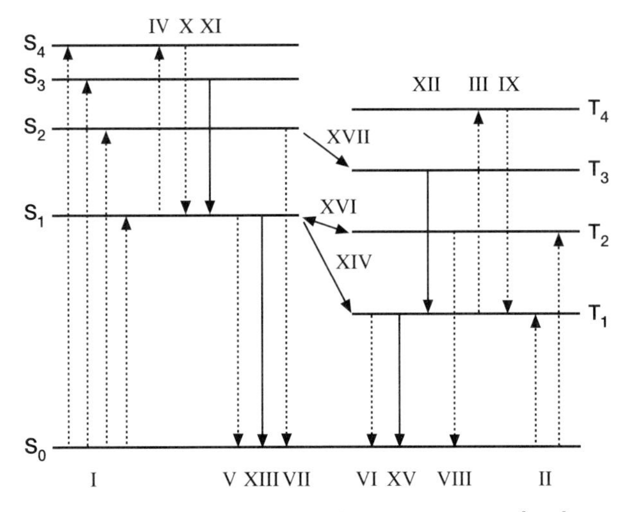

Fig. 9.2. Primary photophysical processes [176]

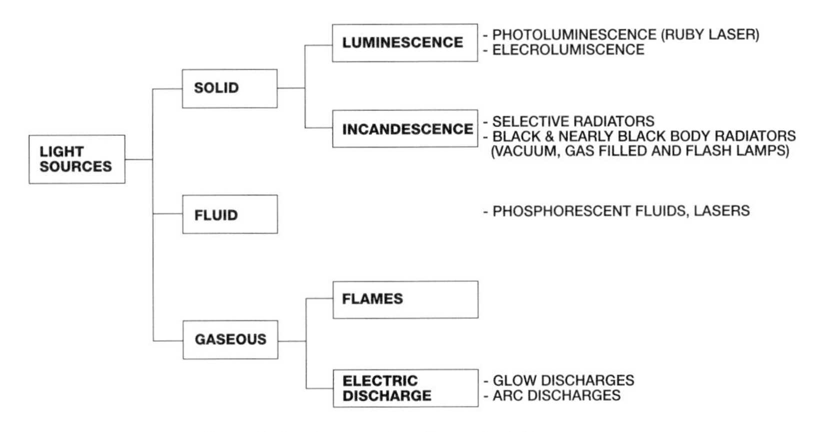

Fig. 9.3. A classification of light sources

also noted that free charges, i.e. those not bound to atoms, radiate when moving nonuniformly.

We may categorise light sources according to how the excitation of atoms and molecules takes place. Typical means of excitation involve the application of an electric current or heat. We may speak, for example, of discharge or incandescent lamps. Also, the dualdivision such as continuum and line sources may be employed to emphasise the spectral properties of the source. Elenbaas [192] has presented a systematic classification of light sources. Its main part is shown in Fig. 9.3.

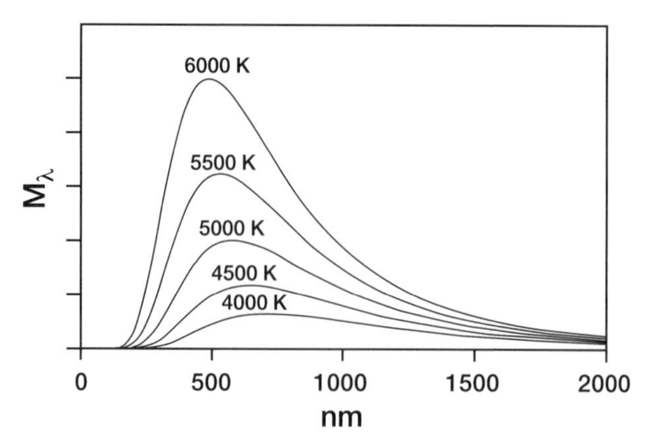

Fig. 9.4. Exitance spectra of a black body at different temperatures calculated using (9.1)

9.1.1 Radiation from a Black Body

It has long been known that material emits electromagnetic radiation as a consequence of its temperature and that this radiation appears over a wide range of the spectrum. At the beginning of the 1900s studies of cavity radiation by the thermodynamics specialist Max Planck (1858–1947) bore fruit. He succeeded in deriving a radiation formula for an ideal emitter, i.e. a *black body.* According to *Planck's radiation law* spectral exitance M_λ is given by [180]

$$M_\lambda = \frac{2\pi hc^2}{\lambda^5}\left(\frac{1}{e^{hc/\lambda kT}-1}\right) , \tag{9.1}$$

where h is Planck's constant, k is Boltzman's constant and T is the absolute temperature of a black body. Figure 9.4 presents the spectra of a black body calculated using the above equation over a temperature range of 4000–6000 K. It is clear from these curves that the radiation maximum moves towards shorter wavelengths as temperature increases. This shift is described in mathematical form by the well-known *Wien's displacement law.*

According to the *Stefan–Boltzman law* the total radiant flux density, in other words, exitance M is proportional to the quadrature of the absolute temperature of the black body, thus

$$M = \int_0^\infty M_\lambda d\lambda = \sigma T^4 . \tag{9.2}$$

In this formula σ has the value $5.670 \cdot 10^{-8}\,\mathrm{W/m^2/K^4}$. At room temperature (20°C) a black body will emit approximately $420\,\mathrm{W/m^2}$. The human eye is capable of detecting radiation generated in this manner provided that the temperature of the body is at least 665 K.

One useful quantity relating to the light source is known as the *colour temperature*. This is the temperature of a black body at which the colours of the black body and the light source correspond with each other. If there is no temperature which produces the same colour sensation then we use the term *correlated colour temperature* to denote the temperature at which the colour of the black body most closely corresponds to that of the light source. Examples of the colour temperatures of a variety of light sources are presented in Table 9.1.

Table 9.1. Approximate color temperatures of various light sources [186]

Source	Color Teperature (K)
Candle flame	1925
Sodium vapor lamp	2200
Electric bulb (100 W)	2890
Tungsten-quartz-iodine	3400
Sun	5300
Carbon arc crater	5500-6000
Xenon lamp	6000
Clear blue sky	15000

9.1.2 Incandescent Sources

A UV/visible light source may thus be constructed by heating a body to a sufficiently high temperature by, e.g. burning it or passing an electric current through it. Light sources which function in this way are described as incandescent.

The functioning of the *halogen lamp* is based on black body radiation. This particularly stable lamp is generally used in the visible and NIR range. Manufactured from tungsten, the life of the halogen lamp filament is made longer than that of its precursor, the tungsten lamp, by the inclusion of some halogen (e.g. iodine or bromine vapour) in the lamp. Thus the lifetime of such a lamp may range from a few tens of hours up to thousands of hours depending on the type of filament. The operational temperature of the lamp will of course also affect its longevity. Tungsten can withstand temperatures of up to 3500°C. Halogens also help to keep the internal surface of the glass bulb clean and so avoid a fall in radiant power resulting from darkening of the glass. Typical electric power ratings of such lamps lie between 10 and 1000 W. Optical power is also affected by the surface area of the filament. Halogen lamps are also available for calibration purposes.

Although halogen lamps do produce thermal radiation they are not usually used as IR sources. These latter include the *Nernst glower*, *Globar* and the *incandescent wire source*. Of these the cylindrically shaped Nernst glower contains rare earth oxides. The temperature of the source is raised to between 1200 and 2200 K with the aid of an electric current. The Globar source is made from silicon carbide and has an operating temperature of approximately 1200 to 1500 K. This source requires water cooling. The incandescent wire source consists of a nichrome metal filament wound into a tight spiral. An electric current is used to heat the filament up to about 1100 K. The lifetime of this IR source is longer than that of the Nernst glower or Globar.

The above-mentioned IR sources differ from one another in their colour temperatures and radiation-producing surface areas. The lifetime of IR emitters is longer than that of halogen lamps, typical figures ranging from thousands of hours to years.

9.1.3 Discharge Lamps

One very common light source for the UV range is the *deuterium lamp*. The gas inside this gas discharge lamp is at a pressure of approximately one hundredth of normal air pressure. The deuterium is excited by applying an electrical current through the gas. As the deuterium molecule returns from its excited state a photon of random energy within the UV range is also produced. This discharge process generates a continual spectrum in the 160 to 400 nm range. In addition to the continuum the Balmer series of D atoms can be seen in visible region. Although this lamp does not produce much visible radiation there is one narrow emission peak of red light at 656.1 nm which is widely exploited for wavelength calibration purposes. In order to maximise the intensity of radiation an aperture is set in between the electrodes inside the lamp. This limits the discharge by restricting the volume of discharging gas to that of a small sphere.

In place of deuterium gas discharge lamps may contain hydrogen or a mixture of the two gases. For operation in the UV range the shells or emission windows of these lamps are made of non-UV absorbing material.

Deuterium lamps and other UV sources produce UV light of considerable intensity. Operators of such lamps should be aware of the safety regulations concerning this region of the spectrum and of the affects of UV radiation on the human physiology. The radiation emitted by light sources operating at wavelengths shorter than 242 nm interacts with oxygen to produce ozone (O_3). While this gas protects life on our planet when high up in the atmosphere it is nevertheless a poison. In small concentrations it irritates the eyes and mucous membranes. UV light has been shown to cause sunburn, pigmentation, lowering of skin elasticity and in the worst cases skin cancer. Powerful UV radiation also causes damage to the eyes. In spite of all this UV radiation also has beneficial effects. In particular, UV–B radiation plays an important role in the synthesis of vitamin D in the skin.

9.1.4 High Pressure Discharge Lamps

When a high radiant power is required over a broad range of the UV/visible spectrum the obvious choice is a gas discharge lamp containing noble gases at high pressure (tens of times greater than atmospheric pressure). The most commonly used noble gas is xenon which produces a relatively even spectrum containing just a few sharp emission peaks mostly in the NIR range. UV emission begins to fall off at about 250 nm and the limit of the useful range lies at about 200 nm. *Xenon lamps* are widely used in fluorescence and luminescence excitation spectroscopy as well as photoacoustic spectroscopy.

Manufacturers produce xenon lamps with electrical power ratings varying from around a hundred watts up to several kilowatts (however, if we wish to examine a single atomic absorption line of width typically 0.001 to 0.005 nm, then a continuum source of even a kilowatt power rating will be insufficient). Xenon lamps are ready to use immediately after being switched on and may therefore be modulated or used as flash lights. The colour temperature of xenon lamps varies between about 5500 to 6000 K. Since this corresponds closely with the colour temperature of the sun xenon lamps may be used in solar simulation.

Discharge lamps may also operate using metal vapour. The most common metal used for such purposes is mercury (Hg). This produces a characteristically spiky spectrum in the UV/visible range. As with xenon lamps these emission peaks rise above a continuous background spectrum. The pressure used in *Hg lamps* varies between 10 and 50 atmospheres. This type of lamp requires a warm-up period before reaching maximum power. Mercury lamps are used as excitation sources in fluorescence spectroscopy.

Inert gases may be used together with metal vapour, one particularly common combination being mercury-xenon. The spectrum of this lamp includes the distinct emission features of the mercury lamp in the UV range as well as the characteristics of xenon emission in the NIR.

9.1.5 Line Light Sources

If we wish to establish the presence of an individual element then our light source must produce an emission line corresponding to some line in the spectrum of the element. This may in principle be achieved by separating out the wavelength in question from a broad spectrum using various filtering techniques or by constructing a light source that actually contains the element in question. The pressure inside a discharge lamp may be lowered to such an extent that the radiation emitted contains only monochromatic wavelengths characteristic of the elements contained in the lamp. For example, at approximately 10^{-5} atmospheres mercury emits radiation almost exclusively of wavelength 253.7 nm. Alternatively such a source may contain sodium (Na), in which case it will produce a yellowish light at 589.3 nm. Ascribed the

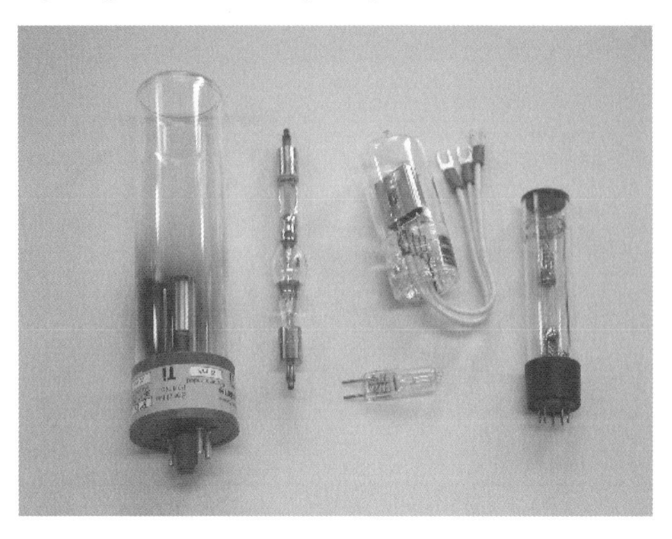

Fig. 9.5. Typical light sources employed in optical spectroscopy. From left to right: a hollow cathode lamp, a xenon lamp, a halogen lamp, a deuterium lamp and a mercury lamp

subindex D, the sodium emission line is used throughout optics as a reference wavelength.[1] Refractive index, for example, is often indicated at this wavelength (n_D).

In the examples presented above light is generated using gas or metal vapour contained inside the lamp. Discharge lamps may also be made such that the required element is present in the solid phase in the cathode of the lamp. One example of such a lamp is the hollow cathode lamp used widely in atomic absorption spectroscopy. As its name suggests a cavity has been machined into the cathode which allows the lamp to produce intense emission.

Radiation emitted from a line light source is not, of course, perfectly monochromatic but covers a narrow wavelength range. The breadth of the band is a consequence of collisions between atoms (known as *Lorentz* and *Holtsmark broadening*), amongst other effects. The *Doppler effect* (inhomogeneous broadening) also serves to broaden the emission line. In the visible spectral range this Doppler broadening is of the order of about 0.001 nm. In addition, the line has its own *natural radiation width* (homogeneous broadening) resulting from small fluctuations in the lifetime of excited atomic states [193]. However, natural radiation width is by no means the first to affect measurements as the other aforementioned mechanisms are of larger magni-

[1] In reality sodium produces two narrow emission lines close to one another, i.e. D_1 (589.6 nm) and D_2 (590.0 nm). These may be distinguished by instruments of sufficient resolution. However, owing to the close proximity of the two lines their average value (589.29 nm) is often used for reference purposes.

tude. Examples of light sources used in spectroscopy are presented in Fig. 9.5. This subject is examined in greater detail e.g. in references [192, 194, 195].

9.2 Controlling Wavelength

An essential part of spectroscopy is the ability of the instrument operator to select and employ those wavelengths which the phenomenon or object to be studied requires. In addition to the selected wavelength the wavelength band is another important parameter in optical spectroscopy. Some applications demand almost perfectly *monochromatic light*[2] while in other experiments a fairly broad waveband is permissible. Proper wavelength is commonly selected by applying one of two main principles. According to the first of these highly monochromatic light is generated simply by using a light source whose optical power is concentrated in a sufficiently narrow wavelength band. Such light sources include lasers and line light lamps. Unfortunately, adjusting the wavelength of these sources is a problematic (or at least expensive) procedure. The second approach is to modify the wide spectral distribution of a light source using optical filters and other external devices. Filters also produce almost fixed reflection or transmission spectra. Spectral devices may also be constructed which contain gratings or dispersion prisms. One advantage these have in comparison with filters is that the nominal wavelength and bandwidth of transmitted light may be adjusted. These devices are known as *monochromators*.

We will now turn our attention to the characteristics and operational principles of optical filters and monochromators designed for use in the UV/visible range. We will concentrate only on those components and methods which effect an essential change in the spectral shape of the incoming light. Thus, filter types such as neutral density filters and colour temperature conversion filters are not considered here.

The technical characteristics of components designed to control wavelength are described by quantities such as transmittance, optical density and wavelength band. Transmittance T is defined in Chap. 8 – in brief it is the relative proportion of intensity I transmitted through the component with respect to the intensity of monochromatic light I_0 arriving at the component. In other words $T = I/I_0$. *Opacity* is the term used to describe the inverse of transmittance $1/T$. The measure of logarithmic transmittance is known as the optical density OD which may be calculated from the transmittance or opacity using

$$OD = \log_{10}(1/T) . \tag{9.3}$$

[2] The common practice of describing electromagnetic radiation covering a narrow wavelength band as being monochromatic is, of course, misleading. Truly monochromatic light is merely a theoretical concept. Light which is almost monochromatic is more accurately described as *quasimonochromatic*.

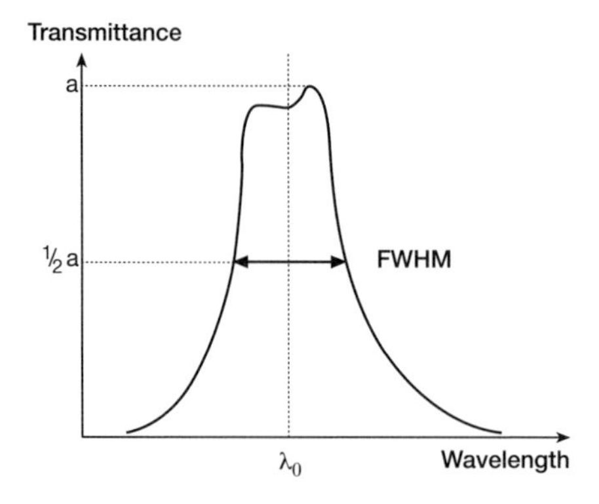

Fig. 9.6. The transmittance curve of an imaginary bandpass filter. FWHM of the curve is defined at half of maximum transmittance, i.e. at $a/2$

This equation may also be used to calculate absorbance A. Furthermore, the concepts of *centre wavelength* and FWHM *bandwidth* (full width at half maximum) are associated with filters designed to select a particular wavelength band, i.e. bandpass from the incoming light. The centre wavelength λ_0 is the wavelength located at the centre of the band while the FWHM bandwidth is the wavelength interval between the two values indicating the half maximum values on the band curve (see Fig. 9.6).

The task of spectral resolution instruments is to distinguish the desired wavelength components from the incoming light. The performance of the various methods available varies considerably. This gives cause for defining the *chromatic resolving power* $\mathcal{R}$ of the system [181]

$$\mathcal{R} = \frac{\lambda_0}{(\triangle\lambda_0)_{\min}} \, . \tag{9.4}$$

In this equation $(\triangle\lambda_0)_{min}$ is the least resolvable wavelength difference at wavelength λ_0. Svanberg [193] defines resolving power in terms of the ratio between the wavelength of monochromatic light and the breadth of the line produced by the spectral instrument at the wavelength in question.

9.2.1 Filters

Coloured Glass Filters

The operation of *coloured glass filters* is based on the ability of colouring matter to selectively absorb different wavelengths. Absorption is effected using simple or complex ions. The most commonly used colouring agents are nickel, cobalt, neodymium, praseodymium and uranium. The minimum effective bandwidth of coloured glass is of the order of 30 nm.

Filters are constructed from coloured glass or coloured gelatine pressed between two pieces of glass. The spectral characteristics of filters are not

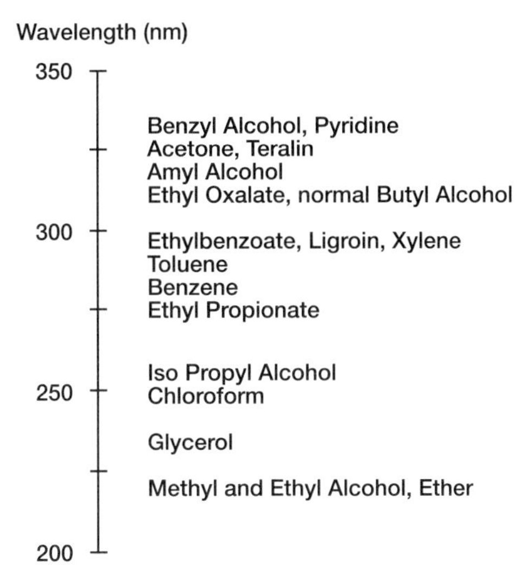

Fig. 9.7. Approximate cut-off wavelengths for selected liquids filters (the location of a liquid with respect to the axis determines the cut-off wavelength)

dependent on the angle of incidence of incoming light and are constant across the entire area of the filter. Similarly, the characteristics of such filters remain virtually unchanged at temperatures below 250°C. Higher temperatures cause permanent changes in characteristics such as transmission.

Coloured glass filters are available for a variety of purposes, e.g. for producing a particular colour sensation or, correspondingly, in order to eliminate some undesirable colour. Coloured glass is very often used to stop transmission completely below a certain wavelength. These filters are known as *cut-off glass filters*. Similarly, coloured glass filters may be used to prevent the propagation of heat, i.e. IR radiation along the light path. However, owing to the nature of the absorption mechanism coloured glass filters are not suitable when high optical power is required. Coloured glass filters are inexpensive compared with interference filters.

Colour filters do not necessarily have to be in the solid phase. The light path may be directed through a quartz glass chamber or cuvette containing some suitable liquid. Organic solutions act as high-pass filters with cut-off wavelengths in the UV area. In Fig. 9.7 we give the approximate cut-off wavelengths of various *liquid filters*. Transmission in only the UV range may be achieved using aqueous solutions of inorganic salts such as a mixture of nickel sulphate and cobalt sulphate. Other compounds commonly used in liquid filters are picric acid, potassium chromate, nitric acid and copper sulphate.

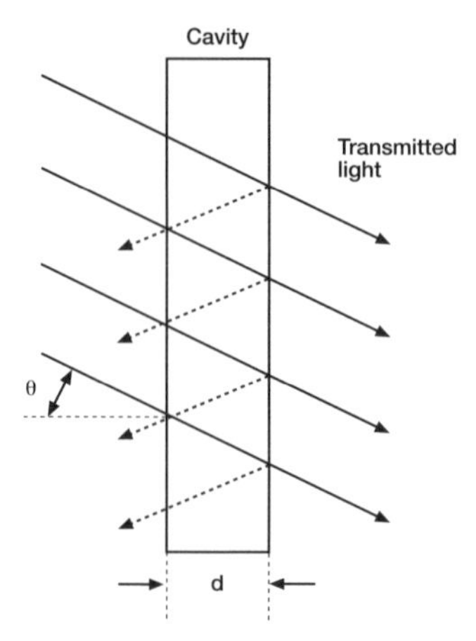

Fig. 9.8. The main concept of an interference filter and a Fabry-Perot etalon is the cavity in which multiple reflection may occur. The wavelengths transmitted depend on the refractive index of the materials involved, the cavity dimensions and the angle of incidence

Interference Filters

The operational principle of interference filters and etalons was elucidated through research performed by C. Fabry and A. Perot in the late 1800s. They were the first to successfully construct a multiple-beam interferometer. The schematic cross-sectional drawing in Fig. 9.8 shows the working principle of such a device i.e. a *Fabry-Perot interferometer.*

Suppose a non-perpendicular collimated light beam from a continuum source arrives at the system. On passing through the first surface boundary the wavefront reaches a space known as the *cavity.* In situations where there is no phase difference between the wavefronts emerging the system the waves reinforce each other. For fixed angle of incidence and cavity dimensions wave reinforcement occurs only at certain wavelengths. Other wavelengths are subject to the destructive interference resulting from the phase difference between the wavefronts. The end result is that only radiation of wavelengths corresponding to the reinforcement case is transmitted through the system while other wavelengths are reflected back towards the light source. Cases of maximum transmission obey the formula

$$m\lambda' = \frac{2d}{\cos\theta} , \tag{9.5}$$

where m is an integer, λ' is the wavelength of radiation in the cavity, θ is the angle of incidence and d is the width of the cavity. The integer m is known as the *order of interference.* Often light arrives at the system almost perpendicularly with the effect that the cosine term on the right hand side of (9.5) equals unity. In this case transmission maxima for wavelength λ are given by $2dn/m$ (where n is the refractive index of the material from which the cavity is constructed). Transmittance falls off rapidly on moving away from these wavelengths. The transmission minimum is obtained when the term m in (9.5) is replaced by $(m + 1/2)$. As transmittance is now at a minimum so reflectance inside the system should correspondingly be at its maximum.

The transmission band may be made extremely narrow and steep by increasing the reflectance of the cavity surfaces. This causes an increase in the resolving power of the filter. Another important characteristic besides the FWHM of each band is the distance between transmission maxima. This is known as the *free spectral range* or FSR. These above-mentioned characteristics together describe the so-called *finesse* F of a filter. This useful parameter is defined by

$$F = \frac{FSR}{FWHM} \,. \tag{9.6}$$

Finesse is affected by the quality of the surfaces, the degree to which they lie parallel to one another and the homogeneity of the materials used.

Air inside the cavity of the Fabry–Perot interferometer may be replaced by a dielectric material, i.e. a transparent insulator which does not possess free charge carriers. Dielectric materials such as calcium fluoride or magnesium fluoride may be used. The width of the cavity must be carefully controlled as the nominal wavelength of the interferometer is dependent on it. In practice the width of the cavity may vary from a fraction of a millimetre to several centimetres (if the system is exploited as a laser resonant cavity considerably greater lengths are used). From an operational standpoint it is important that the walls of the cavity be coated with silver or aluminium to make them highly reflective. In order that undesirable interference is not generated from the outer surfaces of the cavity they are made slightly wedge-shaped.

Both air–spaced and solid *Fabry–Perot etalons* are commercially available. The useable wavelength range covers easily the UV-VIS spectral region, approximately from 190 nm to 2200 nm. Typical values of finesse F are between 25 and 30 according to the manufacturer. The free spectrum range FSR depends on the refractive indices of the materials used and the width of the cavity. In the visible area it is typically between 0.01 and 1 nm. The resolving power of an etalon at zero angle of incidence is given by

$$\mathcal{R} = F\frac{2nd}{\lambda} \,. \tag{9.7}$$

Here n is the refractive index of the cavity. At the same time the free spectral range FSR (measured in wavelength units) may be approximated using the formula

$$FSR = \frac{\lambda^2}{2nd} \,. \tag{9.8}$$

For an imaginary etalon ($F = 30$, $d = 10\,\text{mm}$ and $n = 1$) the above formulae give the following values for resolving power and free spectral range: $R = 1.2 \cdot 10^6$ and $FSR = 0.0125\,\text{nm}$ at a wavelength of 500 nm.

Etalons are used in spectroscopy, diagnostics and laser applications. In spectroscopy they make possible the measurement of hyperfine structure and isotope shifts.

In the realisation of *interference filter* so-called *quarter-wave stacks* are commonly used to enhance reflectance. The effectiveness of the filter may be further improved by placing a number of reflectance cavities – typically between two and four – in series. The transmittance of these *multiple cavity filters* is approximately given by the product of the transmittance of each individual cavity. As a consequence the wings of transmission bands produced by multiple cavity filters are particularly steep and, therefore, the disturbance caused by neighbouring bands is reduced. The cavities are shielded using substrate plates. Undesirable wavelength bands may be eliminated by attaching an additional blocking filter and colour glass to the system.

The transmission band of an interference filter shifts towards shorter wavelengths when the direction of the light beam deviates from the normal incidence. At small values ($0 \ldots 10$ degrees) of θ the centre wavelength λ_o of the displaced band may be calculated using

$$\lambda_o = \lambda_{max} \sqrt{1 - \left(\frac{n_m}{n_e}\right)^2 \sin^2 \theta}\ , \tag{9.9}$$

where λ_{max} is the centre wavelength of the band when the angle of incidence $\theta = 0$, n_m is the refractive index of the medium (usually air) surrounding the filter and n_e is the effective refractive index of the cavity.

Temperature affects both the dimensions of the layers used in the filter and the refractive indices of the materials from which they are constituted. Raising the temperature has the overall effect of shifting the transmission band towards longer wavelengths. In the visible range this shift is of the order of $0.02\,\text{nm}/°\text{C}$. Standard interference filters may be used over a range of temperatures between -50 and $+70°\text{C}$ without any risk of damage. Some special filters are able to withstand temperatures of over 100°C. There is a wide choice of interference filters suitable for the UV/visible range. The bandwidths of commercially available filters vary from 1 to 80 nm.

9.2.2 Monochromators

Man has long known that prisms are capable of dispersing the white light of the sun into its various colour components. Certain colours or wavelengths may then be studied by preventing the propagation of other colours using, e.g. a slit system. The examination of individual colours and choice of wavelength made one particularly important spectroscopic process possible, namely *wavelength scanning*. *Monochromators* repsesent devices in which

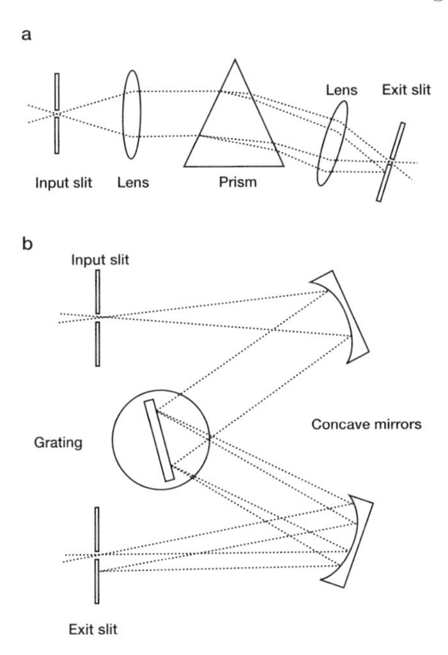

Fig. 9.9. Monochromators employing a different type of resolving component i.e. (**a**) a prism and (**b**) a grating. Light of the desirable wavelength passes through the exit slit while the paths of other wavelengths are blocked

spectrum scanning may be realised. The earliest monochromators functioned using a *dispersion prism*. Later on the prism was almost completely replaced by another light-dispersing component, the *grating* [3]. In both cases the monochromator consisted of a few basic components. In order of the direction of the light path these are e.g. an input slit, a lens or mirror for collimating the light, a component for distinguishing one wavelength from another, a focusing lens or mirror, and finally, at the image plane of the input slit, an output slit. Figure 9.9 depicts the operating principles of a prism (Bunsen) monochromator and a grating (Czerny–Turner) monochromator.

Dispersion Prisms

The ability of a prism to separate white light into its colour components is based on the wavelength dependence of the refractive index of the prism material. This characteristic is known as *dispersion*. Another requirement for

[3] There is also a method relating mainly to the production of IR range spectra by FTIR spectrometers. During measurement the length of one light path of the interferometer in device may be scanned. The data thus collected, i.e. the interferogram is converted into spectral information, e.g. $T = T(\lambda)$, using a Fourier transform calculus.

colour separation is the light refraction i.e. the light enters or leaves the dispersion material or prism at an angle to the normal other than zero. The refractive indices of the most common materials used in optics tend to increase on approaching the UV range from the visible region. Dispersion may be described as the difference between refractive indices measured at differing wavelengths. The *principle dispersion* is defined in the visible range at the wavelengths of the blue (F) and red (C) hydrogen lines (486.1 and 656.3 nm respectively) by the formula $n_F - n_C$. Table 9.2 gives the principle dispersions of some common optical materials.

Table 9.2. Principal dispersion of selected optical material [196, 197]

Material	n_F	n_C	n_F-n_C
Fused silica	1.4631	1.4564	0.0067
Quarzt, e-ray	1.5593	1.5516	0.0077
Quarzt, o-ray	1.5497	1.5419	0.0078
Crown, BK7	1.5224	1.5143	0.0081
Sapphire	1.7756	1.7649	0.0107
Flint, F2	1.6321	1.6150	0.0171

If the width of the base of a dispersion prism is b then its resolving power may be calculated using

$$\mathcal{R} = b\frac{dn}{d\lambda} . \tag{9.10}$$

Here $dn/d\lambda$ describes the change in refractive index with respect to wavelength. A prism made from the flint described in Table 9.2 would yield a value for $dn/d\lambda$ of approximately 1000 cm^{-1} at 570 nm. If the base of a prism made from such material is 5 cm in width then we would obtain a theoretical value for resolving power of 5000. As it is not practicable to build very large prisms the resolving power of prism monochromators is limited.

We note that dispersion depends on the wavelength and increases as we approach the UV range. For this reason the resolving power of a prism is also wavelength dependent. If we wish to keep the band pass of a prism monochromator constant then we need to change the exit slit width as wavelength changes (exit slit width increases as wavelength decreases).

It should be borne in mind when using crystal quartz to make a dispersion prism that this material is birefringent. The polarizing effects of quartz may, however, be cancelled out if the dispersion prism is made by joining together two right-angled triangular (i.e. with angles of 90°, 60° and 30°) quartz prisms, one being of right-handed quartz and the other left-handed. This is known as a *Cornu prism*. One very common prism configuration is that of

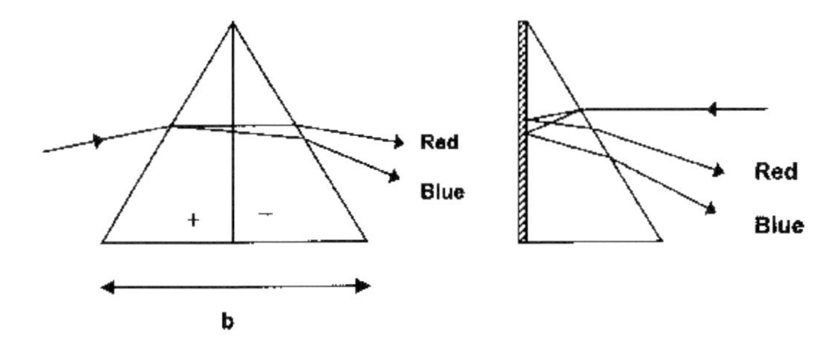

Fig. 9.10. The Cornu prism (left) is made of right and left handed quartz prisms in order to minimise the polarization effect. A Littrow prism (right) goes to form a half of the Cornu prism. Similar functioning is effected by silvering one side of the prism

the *Littrow prism*. Here one side of the right-angled triangular shaped prism is coated by a metal film which acts as an effective reflector. This is used for, e.g. adjusting the wavelength of dye lasers. Cornu and Littrow prisms are presented in Fig. 9.10.

Gratings

Gratings have the same ability to separate light into its various wavelength components as do prisms. However, gratings do not operate according to the principle of dispersion but rather to those of interference and the diffraction of light. The first grating was successfully constructed by J. Fraunhofer in 1823.

A grating is a component, the surface of which contains regular series of narrow adjacent structures. For example, parallel grooves may be cut into an aluminium plate using the sharp point of a diamond. Nowadays commercial gratings are moulded copies of masters and the mechanical processing of gratings has been replaced by other improved methods. For example, the closely spaced interference lines produced by laser beams may be copied into a grating material with help of photoresistant chemicals. The line density of such *holographic gratings* may be as high as thousands of lines per millimetre.

The form and accuracy of the lines in holographic gratings are more homogenous than those of mechanically produced gratings. Neither do holographic gratings produce the unwanted ghost lines characteristic of their mechanical counterparts. Taken together these improvements make possible the generation of cleaner spectra.

Gratings function in either transmission or reflection mode and may be manufactured to operate in different regions of the electomagnetic spectrum. As wavelength grows towards the IR range lower line densities may be employed compared with those required for visible light. Relatively inexpensive gratings of varying sizes are commercially available, the sides of which vary in length from a few centimetres to tens of centimetres.

The operation of a grating in reflection mode may be described by the equation

$$m\lambda = D(\sin\theta_i + \sin\theta_r) , \tag{9.11}$$

where D is the groove separation of the grating, θ_i and θ_r are the angles of incidence and reflection respectively, and m is an integer, i.e. the order of diffraction. The resolving power of a grating depends on the number N of illuminated lines and the order of diffraction m. Thus resolving power is given by

$$\mathcal{R} = Nm . \tag{9.12}$$

For example, if a 5×5 cm grating possessing 1000 lines/mm is completely illuminated its resolving power is 50000 in diffraction of the first order.

On examining the formula 9.11 we notice that at a given orientation θ_r other wavelengths besides the desired wavelength λ may also be reflected. Indeed, higher orders, i.e. wavelengths of $\frac{1}{2}\lambda$, $\frac{1}{3}\lambda$ etc. may also be diffracted in the same direction. These generally undesirable components may be easily eliminated by using a suitable filter or two monochromators in series.

The *blaze wavelength* indicates the wavelength at which *grating efficiency* is at its highest. Best figures for grating efficiency may be as high as 70–80%. Unfortunately, grating efficiency is high only for a relatively limited wavelength range. However, this spectral range may be controlled in a manufacturing process (using groove shaping techniques, or blazing). Efficiency is also highly dependent on the polarisation of incident light with respect to the orientation of grooves on a grating.

With traditional monochromators only the first or other low orders of diffraction are employed. However, R.G. Harrison's *echelle gratings* first presented in 1949 are designed to operate at high orders of diffraction. For this reason they are capable of high resolving powers, sometimes exceeding even 10^6. In such cases the angle of incidence must also be high.

Established grating monochromator geometries include Ebert–Fastie, Czerny–Turner and Sey-Namioka [193].

9.3 Light Polarization and Polarizing Components

The essence and behaviour of light may be visualised in terms of the theory of wave optics. It has been shown that light is a form of transverse wave motion in which electric and magnetic fields oscillate in directions perpendicular to that of light propagation as well as mutually perpendicular to one another. An essential feature of this transverse wave is the concept of *polarization*. Indeed, the elaborate control of light polarization has proved an invaluable resource in numerous measurement applications.

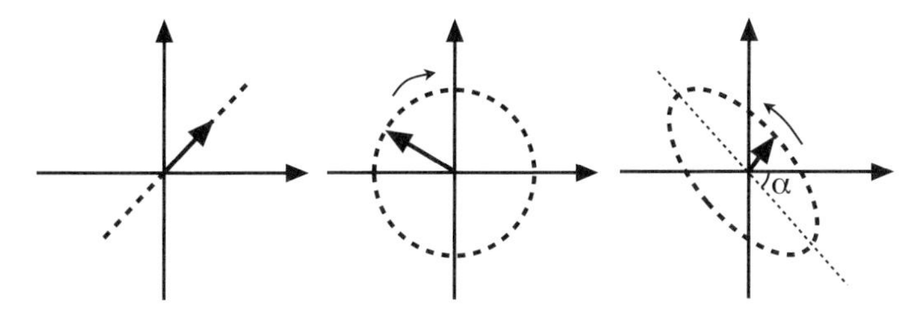

Fig. 9.11. The motion of the electric field of a wave may be pictured by a vector. Here the tip of the vector plots a straight line, circle and ellipse, corresponding to linear, circular and elliptical polarizations, respectively. The rotation direction of a vector (if any) with respect to time offers an additional attribute to polarization. Therefore the case in the middle represents a right-circularly polarized wave providing that the wave is traversing toward the observer

If our sense of vision were sufficiently developed to allow us to directly observe the oscillations of electric and magnetic fields then we could study polarization by performing the following experiment. In this experiment we would observe the light approaching us from the light source and would pay particular attention to the directions of oscillation of the two fields. To aid observation we would use two vectors describing both the time-dependent magnitudes and directions of the oscillations. Because we were examining only a single possible light path the tails of the vectors would be located in the path. Furthermore, the observation vectors would be positioned at a fixed distance between ourselves and the light source. How the vectors behaved would depend on our light source. In some cases we would notice that the vectors would present themselves only in fixed directions or that the tips of the vectors would appear to describe circular or elliptical paths – the form of the light being in some respect regular as shown in Fig. 9.11. In such cases we would be dealing with *polarized light*. With a different light source observations might show the directions of the vectors to vary at random, i.e. not according to any regular pattern. Such a light source produces *unpolarized light*.

Although the human eye is almost incapable of directly observing the polarization of light[4] polarization techniques and their applications have established a firm foothold in many areas of science and industry such as astrophysics and molecular biology, to name just two. The polarization of light transmitted through a sample may change as a result of e.g. mechanical stress or chemical reactions occurring within the sample. Polarization techniques are particularly useful in the study of photochemical and photobiological processes. In addition, the study of layered structures relies for the main part on polarization. For a more detailed examination of the fields of ellipsometry and

[4] We refer the reader to the fascinating section "Polarization blindness can be cured" on p. 54 of [201].

polarization spectroscopy we refer the reader to references [198–201]. Considering this theme from a practical perspective we will concentrate on the general nature of light polarization, the mechanism by which it is generated and the components and devices with which polarization may be produced and controlled.

9.3.1 Characteristics of Light Polarization

We may picture the sun or some incandescent lamp filament as being made up of a large number of randomly oriented emitters. On returning from the excited state the atoms or molecules of the source each emit a short (approx. 10 ns) and polarized wavetrain. During this instant all wavetrains emitted sharing the same frequency combine to form a single polarized wave. This process is random, extremely quick and continuous. As a consequence of this we are unable to distinguish any dominant polarization state in the radiation emitted from the source. We call such light *natural light.* The slightly misleading term *unpolarized light* is also widely used [181].

Polarization States

The generation of polarization states may be conveniently investigated using the formulae governing the propagation of light. Maxwell's concept of light is based on electric and magnetic fields oscillating at the same frequency. The two fields lie perpendicular with respect both to one another and also to the direction of light propagation. The effects of the light on the atoms and molecules are mostly related to the rearrangement of charges. The electric field plays the leading role in this process while the effect of the magnetic field is negligible. Here, we too may disregard the magnetic field and are free to consider waves exclusively in terms of the electric field. One of the most fundamental principles of wave motion is known as the *Principle of Superposition* [202]. According to this we may deduce what will take place when two or more arbitrary waves meet one another. We will encounter examples of this phenomenon when we now turn our attention towards the various polarization states.

Consider two waves sharing the same frequency ω propagating in the direction of the z axis. Let us further suppose that the electric fields of these waves perform planar oscillation in the directions of the x- and y-axes, respectively – in other words, both waves are polarized. The time- and space-dependent electric fields may be presented in the form

$$\begin{aligned} \mathbf{E}_x(z,t) &= \hat{\mathbf{i}} E_{0x} \cos(kz - \omega t) \\ \mathbf{E}_y(z,t) &= \hat{\mathbf{j}} E_{0y} \cos(kz - \omega t + \phi)\,, \end{aligned} \tag{9.13}$$

where ϕ is the relative phase difference between the two waves, E_{0x} and E_{0y} are their respective amplitudes and $k(= 2\pi/\lambda)$ is the wavenumber. The oscillation of the electric field is represented here by the cosine function. It should be noted that propagating waves are also commonly described using complex presentation. Since both waves oscillate in planes we may speak of *linearly polarized* or, alternatively, *plane-polarized light.* According to the principle of superposition the resultant wave may be obtained from the vector sum of these two individual waves

$$\mathbf{E}(z,t) = \mathbf{E}_x(z,t) + \mathbf{E}_y(z,t) \; . \tag{9.14}$$

If our waves are in phase, i.e. $\phi = 0$ we also obtain

$$\mathbf{E} = (\hat{\mathbf{i}}E_{0x} + \hat{\mathbf{j}}E_{0y}) \cos{(kz - \omega t)} \; . \tag{9.15}$$

From (9.15) we notice that the resultant wave is also linearly polarized but that it oscillates in a plane which lies in between the x- and y-axes. The precise orientation of the plane of oscillation is determined by the amplitudes E_{0x} and E_{0y}. The relative phase difference $\pm\pi$ also produces a plane-polarized resultant wave.

Two plane-polarized waves may be combined into a single plane-polarized resultant wave. This phenomenon may also be considered in reverse order. It may be shown that every arbitrary plane-polarized wave may be separated into two orthogonal, plane-polarized components. Light which is linearly polarized is said to occupy the *P-state.*

We shall now consider a situation in which the phase difference between the two waves is $-\pi/2$ and their amplitudes are identical, i.e. $E_0 = E_{0x} = E_{0y}$. Thus we may calculate the resultant vector

$$\mathbf{E} = E_0 \left[\hat{\mathbf{i}} \cos{(kz - \omega t)} + \hat{\mathbf{j}} \sin{(kz - \omega t)}\right] \; . \tag{9.16}$$

We note that the resultant vector has a constant amplitude E_0. The direction of the E vector does not, however, remain fixed but varies through time. Since E_0 is constant the tip of the vector describes a circle as shown in Fig. 9.11. This form of polarization is therefore known as *circular polarization.* The rotation of the vector naturally has a direction. In this case the vector appears to rotate clockwise when viewed from the direction towards which the wave is propagating. The polarization state may be defined using this information. The wave in the present case is described as being *right-circularly polarized.* If the phase difference is $+\pi/2$ then the sign in front of the unit vector $\boldsymbol{j}$ in (9.16) changes into a negative. We would then have *left-circularly polarized light.*

As a consequence of its direction of rotation circular polarized light occupies either the *R-state* or *L-state.* Two circular polarized waves with differing directions of rotation but identical amplitude E_0 combine to generate a linearly polarized wave. The amplitude of this new wave is $2E_0$.

When an electromagnetic wave encounters matter it may lose both energy and linear momentum. In the case of circular polarized light we may assume that the rotating electric field of the wave causes the circular motion of electrons. In so doing the wave also transfers angular momentum to the material.

We have still to consider one more important form of polarization, namely, *elliptical polarization.* Its importance lies in the fact that linear and circular polarization may both be regarded as special cases of elliptical polarization. Here we no longer set restrictions concerning the respective amplitudes of or relative phase difference between the two waves. When two waves of this kind are combined the result is a wave, the electric field vector of which describes an ellipse. By examining only the magnitudes of the waves in question it may be shown that [181, 203]

$$\left(\frac{E_x}{E_{0x}}\right)^2 + \left(\frac{E_y}{E_{0y}}\right)^2 - 2\left(\frac{E_x}{E_{0x}}\right)\left(\frac{E_y}{E_{0y}}\right)\cos\phi = \sin^2\phi\,. \tag{9.17}$$

This equation describes an ellipse as shown in Fig. 9.11. In the figure the angle α between the major axis of the ellipse and x axis may be calculated using

$$\tan 2\alpha = \frac{2E_{0x}E_{0y}\cos\phi}{E_{0x}^2 + E_{0y}^2}\,. \tag{9.18}$$

As in the case of circular polarization the phase difference ϕ determines the direction of rotation of the vector. Elliptically polarized light is said to exist in the *E-state.*

It should be noted that phase differences may also be multiples of wavelength or 2π. For example, in the case of linear polarization the phase difference may be written in the general form $\phi = \pm 2\pi m$ where m is an integer. This naturaly also applies when $\phi = 0$.

Stokes' Polarization Parameters

In the middle of the 1800s G.G. Stokes (1819–1903) came to a significant conclusion, namely, that all polarization states may be described by four parameters. These *Stokes' polarization parameters*, $S_0 \ldots S_3$, are of great practical value as they may be calculated directly from intensity measurements. We have already perceived that all polarization states may be explained in terms of the mutual interaction between two linearly polarized waves. Using the notation of (9.13) the Stokes' parameters for the resultant wave take the form

$$\begin{aligned}
S_0 &= E_{0x}^2 + E_{0y}^2 \\
S_1 &= E_{0x}^2 - E_{0y}^2 \\
S_2 &= 2E_{0x}E_{0y}\cos(\phi_y - \phi_x) \\
S_3 &= 2E_{0x}E_{0y}\sin(\phi_y - \phi_x)\,.
\end{aligned} \tag{9.19}$$

The first parameter S_0 describes the total intensity of the optical field. The remaining three parameters describe different polarization states, i.e. S_1 represents the vertically and horizontally oriented components of linear polarization, S_2 represents linear polarization directed along $\pm 45°$ and S_3 the component of polarization rotated to either the left or right.

The classical method of determining these parameters employs two polarization elements, i.e. a retarder and a polarizer, in series. This procedure is relatively quick, involving just four measurements performed at differing relative orientation of the two elements. Details of this method are presented e.g. in reference [199].

Degree of Polarization

Natural light and completely polarized light constitute the two extreme cases on a scale indicating the relative proportions of the various polarization states. The *degree of polarization* P is defined as the ratio of intensity of all polarization components I_{pol} to total intensity I_{tot}. The degree of polarization may be calculated using the Stokes' parameters

$$P = \frac{I_{pol}}{I_{tot}} = \frac{\left(S_1^2 + S_2^2 + S_3^2\right)^{1/2}}{S_0} . \tag{9.20}$$

The intensity of the polarization components of natural light is zero, therefore $P = 0$. Correspondingly, in the case of totally polarized light $P = 1$. In practice all the light we observe and use is *partially polarized*.

9.3.2 Mechanisms by which Polarization is Produced

There are currently available numerous devices designed to produce and control polarization. They are all based, however, on one of four basic concepts. These are birefringence, dichroism, reflection and scattering.

Birefringence

The first written reports of *birefringence* date from the 1600s. They concern the observation that certain naturally occurring transparent crystals divide incident light into two separate beams as shown in Fig. 9.12. Later on it was noticed that one of these beams obeys Snell's refraction law while the other does not. These became known as *ordinary rays* (o-rays/waves) and *extraordinary rays* (e-rays/waves). For a long time birefringence remained a mystery and it was not until 1817 that T. Young came up with an explanation of the phenomenon. His answer was based, quite expectedly, on polarization.

Unlike liquids and amorphous solids, many crystals exhibit anisotropic features. The structures of these crystals are often hexagonal, tetragonal or

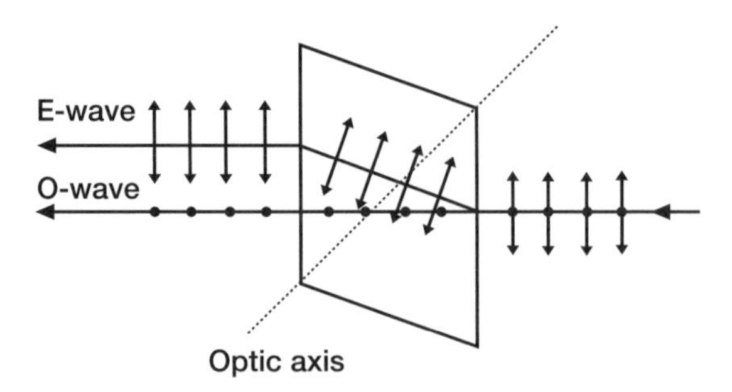

Fig. 9.12. A correctly cut and polished birefringent crystal divides the incoming wave into two orthogonal plane-polarized components

trigonal. This means that owing to its regular atomic structure the optical properties of such a crystal are dependent on the direction of observation. In cases of a *uniaxial crystal* structure there will be one particular direction at which light separation does not occur. This direction is known as the *optic axis* of the crystal. Light polarized in the direction of the optic axis, i.e. symmetry axis, (extraordinary rays) propagates at a speed proportional to the refractive index of the material n_e. Correspondingly, light polarized in a direction perpendicular to the symmetry axis (ordinary rays) will experience a refractive index of n_o. Such crystals may be used in polarization measurements. Indeed, an accurately cut crystal will, for example, separate the wave into two orthogonal, plane-polarized waves.

There are also crystals with two optic axes. These *biaxial structures* may be orthorhombic, monoclinic or triclinic. One example of such a structure is *mica* which is a widely used material in constructing wave plates.

An example of a naturally occurring uniaxial crystal used in optics is *calcite*, a rhombohedral crystalline form of calcium carbonate. Largish calcite crystals may still be found today, especially in parts of India, Mexico and South Africa. For this material $n_e < n_o$ and so calcite is said to be *optically negative*. Although there are numerous other optically negative and positive materials to choose from calcite is perhaps the most commonly used in birefringence-based linear polarizers. Dispersion curves of both optically negative and positive materials, namely calcite and quarzt are shown in Figs. 9.13 and 9.14. Properties of various crystals may be found e.g. in [197, 204–206].

The advantages of calcite are its availability in the desired size, its transparency, non-hygroscopicity, non-toxicity and the ease with which it may be cut and polished. Although there are no generally agreed quality criteria concerning commercial calcite the selection of suitable material should take into account its colour as well as its tendency to diffuse light and cause wavefront distortion or striae.

Calcite may be used over a wide range of the spectrum (260 to 1700 nm). Although o-rays have a higher transmission they are often ignored in such

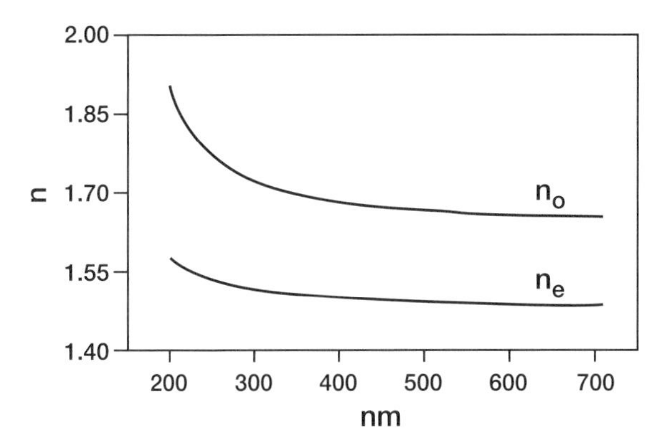

Fig. 9.13. Natural calcite is an optically negative anisotropic material. To describe its optical behaviour two refractive indices, n_e and n_o, are required

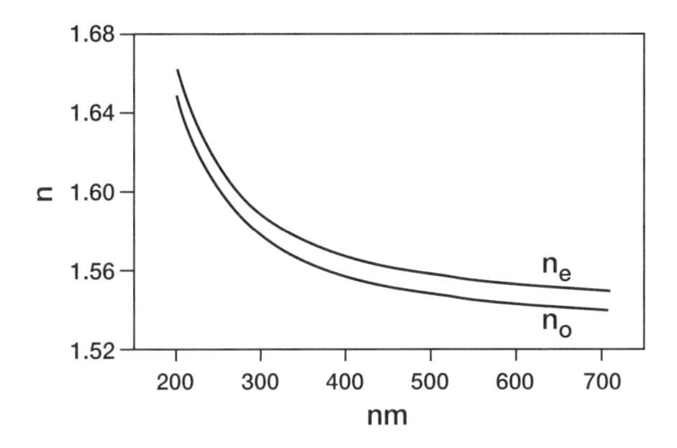

Fig. 9.14. Dispersion of an optically positive crystal, i.e. quartz

polarizers. Because origin of polarization here is not based on absorption these materials may be used in high power laser applications. Other common birefringent crystals include tourmaline, quartz, sodium nitrate, ice and rutile (TiO_2).

Dichroism

Dichroism concerns the selective ability of matter to absorb electromagnetic radiation in specific directions with respect to the orientation of the matter. In other words, absorption only effects one of the orthogonal components in the P-state while to the other component the matter is transparent. One such naturally occurring substance is anisotropic tourmaline. The electric field components which are perpendicular to the optic axis of the crystal are effectively absorbed by the material (see Fig. 9.15). Consequently, a sufficiently thick tourmaline crystal will completely eliminate the other polariza-

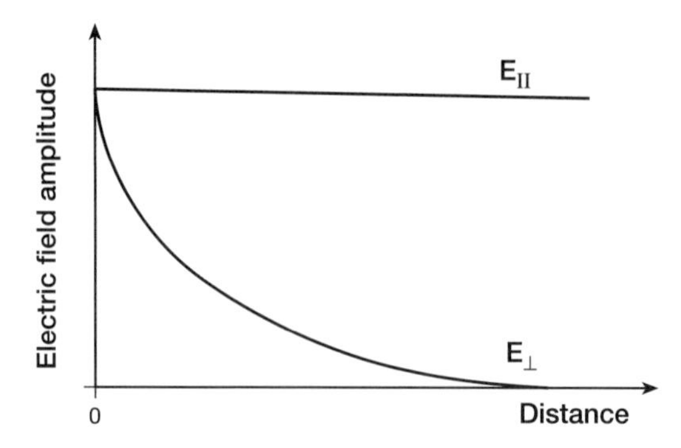

Fig. 9.15. Dichroism refers to the selective absorption of material. In a linear polarizer made from such a crystal, e.g. from tourmaline, the electric field perpendicular to the transmission axis of the polarizer is absorbed while the other orthogonal wave travels through the system essentially unattenuated

tion component. The usefulness of tourmaline is limited by the wavelength dependence of absorption and the relatively small size of naturally occurring crystals.

Dissipative absorption relates to the conversion of electromagnetic field energy in a material into joule heating. This occurs when atoms in the excited state collide with one another or when the charges in the material which are set into motion by the electric field lose their energy through friction. Metals are known to contain large quantities of free charge carriers and are for this reason good electrical conductors. The generation of electric currents in a specified direction, i.e. selective absorption, may be effected if we construct a system in which thin metal wires are placed both parallel and very close to each other. Those electric field components lying in the same direction as the wires generate in them a macroscopic oscillating current. This current in turn gives rise to joule heating. Absorption does not, however, occur in the direction perpendicular to the wires. This direction therefore defines the *transmission axis* of the system (see Fig. 9.16).

Polarizers built according to this principle are known as *wire-grid polarizers*. Owing to problems in the manufacturing process it is not practicable to build such kinds of devices for use in the visible range. The principle may, however, be exploited with IR and longer wavelengths (e.g. Hertz's experiments polarizing radio waves).

The dichroism-based polarizer most widely favoured today came into being as a result of the research and innovations of E.H. Land in the early 1900s. Nowadays his inventions are known by the trade names *Polaroid J-sheet* and *H-sheet*. Especially the latter of these is an extremely common component in the generation of linearly polarized light. In H-sheet manufacture sheet polyvinyl alcohol is rendered anisotropic by a stretching process. The stretched sheet is then made into an electric conductor through iodine

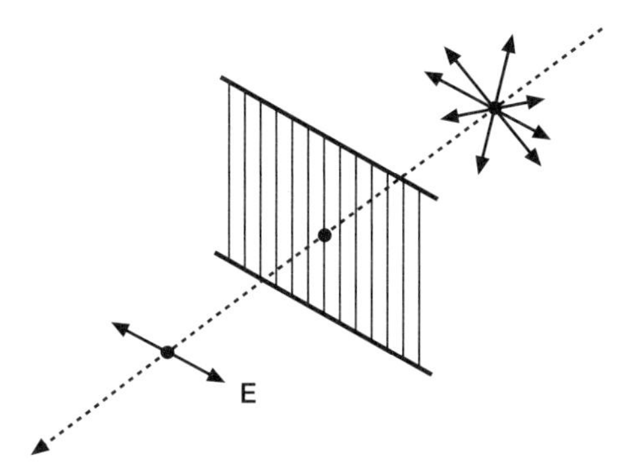

Fig. 9.16. A wire-grid polarizer consists of numerous thin and closely spaced conductors. The transmission axis of the polarizer is perpendicular to the wires

treatment, i.e. it is soaked in a concentrated iodine solution during which the iodine attaches itself to the oriented polymeric molecules in the sheet. Thus, electric conductors are fixed parallel to one another on the sheet in much the same manner as in a wire-grid polarizer. Since the manufacturing process of the component now involves changes at the molecular level the line density of the electric conductors is therefore high and the component may thus be used in the visible range.

Reflection

Let us consider a ray of light arriving at a glass plate as shown in Fig. 9.17. At a certain angle of incidence we notice that the light reflected from the surface of the glass is linearly polarized [207]. Here the electric field of the linearly polarized wave oscillates parallel with respect to the boundary surface (s-polarization). At the same time, the refracted wave is only partly polarized. This effect of light reflection and polarization was studied at the beginning of the 1800s by E. Malus and D. Brewster, amongst others. Brewster came to the empirical conclusion that the degree of polarization of reflection is at its greatest when the angle between the reflected and refracted light rays is 90°.

The angle of incidence, or, alternatively, angle of reflection, at which total polarization occurs is known as the *polarizing angle* or *Brewster's angle* θ_B. The polarization angle may be calculated if we know the refractive indices n_1 and n_2 of the two media. By applying *Snell's law* we obtain

$$\tan\theta_B = n_2/n_1 \; . \tag{9.21}$$

In an air-glass system the polarization angle is approximately 57°. Light polarization as a result of reflection is, however, relatively inefficient because only some 10% of incident light reflects at the polarization angle. The remainder is refracted in a beam which is only partly polarized. It is worth

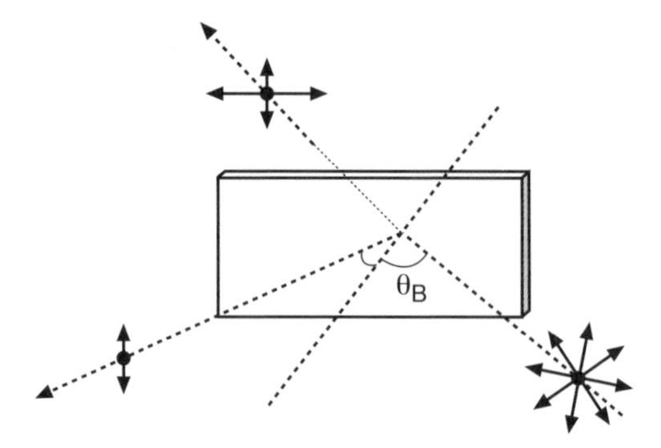

Fig. 9.17. When the incident light encounters the glass plate at a specific angle, i.e. the Brewster angle θ_B, the reflected light is plane polarized

noting that light reflected from angles of incidence other than the polarization angle is also polarized but only partly so. Since the reflection of light in nature is such a commonplace phenomenon so partly polarized light may be found all around us.

Scattering

When a gas at low pressure is exposed to unpolarized light the electrons of its molecules begin to vibrate in directions perpendicular to that of the propagating wave. In this way the energy from the optical field of the molecules reverts to dipole radiation. A dipole may emit radiation in all directions except that of its own axis. Consequently, an observer looking perpendicularly at the propagating beam will notice that the scattered light coming towards him is plane-polarized. This plane of vibration is always perpendicular to the direction of propagation of the original wave.

This phenomenon may be observed by examining the sky. Light from the sun is scattered and polarized by the molecules of the atmosphere. This may be studied using a linear polarizer – the proportion of transmitted light is clearly dependent on the orientation of the polarizer. Polarization in the atmosphere is not, however, total owing to multiple scattering and the presence of large atmospheric particles.

9.3.3 Polarization Components

Defined in general terms, a polarizer is a component which changes the polarization state of the incoming light, which may itself be natural light or already polarized. Nowadays there are numerous commercially available polarizers intended for a wide range of applications. When choosing a polarizer attention should be paid to its operational selectivity and accuracy, effective wavelength range, measurement geometry, the light source power and the efficiency of the whole device. Some polarizers are particularly expensive so, naturally, price also has to be taken into consideration.

In planning the experimental arrangement it should be noted that many other components besides polarizers cause some degree of light polarization. The monochromator with its slits, gratings and mirrors is a good example of a device which polarizes the light passing through it in a more or less complicated manner.

Let us now consider the most important components employed in polarization measurements. These are the *linear polarizer* and the *retarder*. When an experiment requires the use of unpolarized light we may synthesise it from polarized light using a *depolarizer*. Here we will also briefly examine depolarization methods.

Linear Polarizers

The purpose of the linear polarizer is to convert the light directed at it into a form of light the electric field of which oscillates in a particular chosen plane. Those electric field components which lie perpendicular to the transmittance axis of the device are absorbed, reflected or deflected from their original direction by the mechanisms described in previous section. The operator may choose at will the direction of the oscillation plane by simply turning the polarizer in relation to the axis of the light beam. When two such (ideal) components are placed in series the intensity of transmitted light may be calculated using

$$I(\theta) = I_{\|} \cos^2 \theta \, . \tag{9.22}$$

Here $I_{\|}$ is the transmitted intensity when the transmission axes of the polarizers lie in the same direction. This formula is known as *Malus' law*. If the transmission axes are *crossed*, i.e. at right angles to one another, the light ray will be completely absorbed by the second polarizer (also known as the *analyzer*). In practice the complete extinguishing of the light beam does not happen and a small part passes through the system. This deficiency is described by the *extinction ratio* H

$$H = \frac{I_{\perp}}{I_{\|}} \, , \tag{9.23}$$

where $I_{\perp}$ is the intensity of transmitted light for the case in which the transmission axes lie mutually perpendicular. Typically the extinction ratio takes values ranging from 10^{-2} to 10^{-7} depending on the accuracy of manufacture and chosen polarization mechanism. It should be noted that H is generally wavelength dependent.

Linear polarization may be achieved in a number of ways. More specialised alternatives include the placing of glass plates at the Brewster angle and the use of polarizers which exploit light scattering. These are used in applications where common commercial polarizers fail to produce satisfactory results, such

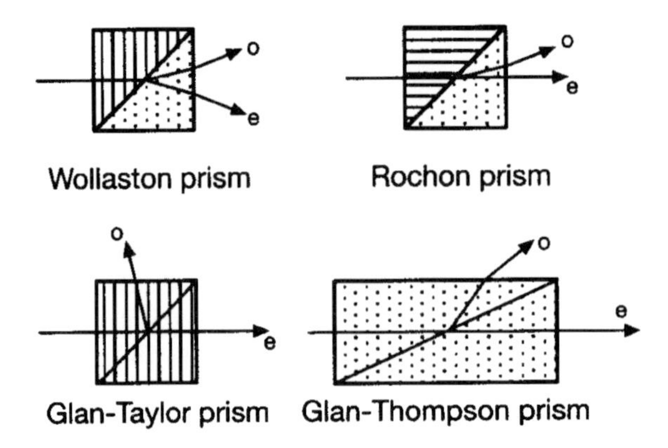

Fig. 9.18. Typical prism polarizer configurations. Wollaston and Rochon prisms function according to the principle of birefringence. The direction of the optic axis is denoted by dots and lines. In Glan-type prisms o-waves are redirected by the effect of total reflection

as in the vacuum-ultraviolet range or in more remote regions of the infrared. We will leave these less common methods behind us and concentrate on the two types of polarizers most widely used in optics. These are the *prism polarizer* and the *sheet polarizer*.

Prism polarizers operate according to the principles of birefringence or total internal reflection. They are made from two prisms fixed together. Earlier prism polarizers were the only practicable polarizing devices available. Nowadays, with the exception of certain specific applications the prism polarizer has been effectively superseded by sheet polarizers.

However, the prism polarizer is the best alternative for work in the UV range. The lower limit of the range is determined by the prism material and the way in which the prisms are attached together. If the prisms are fixed using an optical cement such as Canada balsam the available wavelength range will extend as far as approximately 330 nm. If, on the other hand, the prisms are arranged with an air gap in between then the useful spectral range will depend exclusively on the prism material. Calcite, for example, begins to absorb at around 240 nm while its useable range extends a few tens of nanometres below this. We may delve still deeper into the UV range by using prisms made from magnesium fluoride or crystalline quartz.

Prism polarizers come in a number of basic types. The most widely known of these include the *Glan–Taylor*, *Glan–Thompson*, *Wollaston* and *Rochon prisms* (see Fig. 9.18). These differ from one another in the manner in which the polarized rays are directed out from the polarizer. In Glan-type polarizers (made by calcite or other optically negative material) one of the polarized components, i.e. the ordinary waves, experiences total reflection at boundary surface between the two prisms while the other polarized component, i.e. the extraordinary rays, is transmitted through the system. This ingenious method of generating linearly polarized light works by exploiting the difference in refractive index between the different rays. In Wollaston and Rochon prisms both polarization components emerge from the same side of the polarizer but at slightly different angles.

Common to all prism polarizers are good values for the extinction ratio H, i.e. typically better than 10^{-5}. On the other hand, polarizers based on total reflection only function well over a narrow, wavelength-dependent angle of incidence. With Glan–Taylor polarizers the *acceptance angle*, i.e. full-field angle of rays for which the polarizer is effective, is about $5 \ldots 10$ degrees while for Glan-Thomson devices it is of the order of $15 \ldots 25$ degrees depending on the dimensions of the polarizer. Prism polarizers are also relatively bulky with input areas of 10×10 or 20×20 mm and typical lengths of several centimetres.

Prisms can operate at very high light intensities providing that the rejected beam is directed away from the polarizer via separate exit port. Also, prisms for use with high intensities may not be fixed together with cement. Developed for use in laser applications, the *Glan-laser* polarizer achieves a transmittance of around 90% for a polarized light input. This device has a good extinction coefficient while its acceptance angle is just a few degrees in the visible range.

Sheet polarizers take advantage of the phenomenon of dichroism. As mentioned above, modern commercial sheet polarizers are based on the pioneering research of E.H. Land and, in particular, the development of the H-sheet.

This type of polarizer functions well in the visible range with extinction ratios reaching 10^{-5} (see Fig. 9.19). Often the extinction ratio is limited by protective glass plates placed on either side of the active sheet. Since the polarization changes as a result of mechanical stress so the glass plates must be stress-free. In the UV range glass is replaced by fused silica plates. However, for work in the UV range sheet polarizers are not so attractive a choice as prism polarizers because of their poor extinction ratios (10^{-2}). Sheet polarizers are also available for use in the NIR range although here too their level of performance falls behind those designed for the visible range.

Owing to selective absorption sheet polarizers cannot be used in conjunction with high-powered lasers. Typical applications include the analysis of polarization states, electro-optical and magneto-optical modulators, and the elimination of reflection.

Although sheet polarizers cannot achieve the same extinction ratios over a wide range of the spectrum as prism polarizers they, nevertheless, possess other excellent and useful properties. Their useable acceptance angle is wide, exceeding 30 degrees in the visible range. With respect to the direction of light propagation sheet polarizers are just a few millimetres thick, inclusive of their protective glass plates. The clear aperture of sheet polarizers may

Fig. 9.19. When two polarizers are crossed essentially no light passes through the system. The transmission axes of these sheet polarizers are indicated by white lines engraved on their frames

also be made larger than those of prism polarizers – apertures as large as 100 mm are easily available. Finally, their relatively low cost has helped to increase their popularity.

Retarders

The purpose of a *retarder* (or a *wave plate*) is to effect a phase difference between the two orthogonal polarization components such that on emission the polarization of the resultant wave is altered. The retarder may therefore be used to change one given polarization state into another. The operational principle of the retarder is important because it is capable of producing both circular and elliptical polarization. The desired polarization may be achieved at a suitable phase difference. This in turn depends on the structure and material of the retarder.

The phase difference of a retarder may also be adjustable. Such retarders are known as *compensators.* The phase difference brought about by non-adjustable retarders is in general expressed as some fraction of a wavelength, such as $\frac{1}{4}\lambda$ or $\frac{1}{2}\lambda$. Units of length may also be used. Thus a 150 nm retarder denotes a quarter-wave retarder at a wavelength of 600 nm. The orthogonal polarization components arriving at the retarder may consist of linearly, circularly or elliptically polarized waves, although the latter two are less commonly used.

The functioning of the retarder is usually based either on the characteristics of birefringent materials or on wave reflection. Here we will concentrate

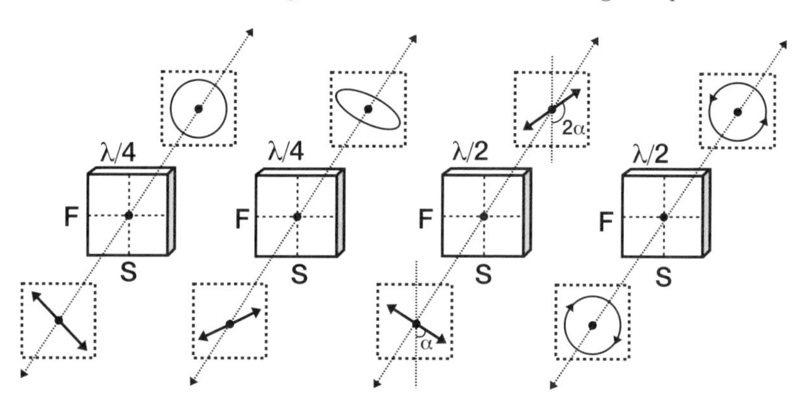

Fig. 9.20. A quarter-wave plate and a half-wave plate change the incident polarization. F and S denote the fast and the slow axes of the plate

on retarders made from calcite. A crystal of calcite may be cut and polished in such a way that its front and reverse surfaces are parallel and that the optic axis of the crystal also lies in the plane parallel to these two surfaces. When a wave arrives at the crystal it splits into two parallel waves (one o-wave and one e-wave). Since the birefringence ($\triangle n = n_e - n_o$) of calcite is negative so the o-wave remains behind the e-wave. In the case of calcite and other optically negative uniaxial retarders the direction described by the optic axis is known as the *fast axis* while the direction perpendicular to it is called the *slow axis*. Since the electric fields of the waves vibrate at right angles to one another and the waves are of the same frequency so on leaving the crystal the two waves combine. If the waves travel a distance d in the crystal at different speeds then a phase difference $\triangle\varphi$ will be generated

$$\triangle\varphi = \frac{2\pi}{\lambda} d |\triangle n| \,. \tag{9.24}$$

Here λ is the wavelength in vacuum. This phase difference is known as the *retardance* or *retardation*. It should be noted that if plane-polarized light arriving at the retarder is parallel with either the fast or slow axis then no phase difference can be generated as one of the two required components will be absent. As a result the polarization state of the light will remain unchanged.

Since the difference in refractive indices $\triangle n$ is virtually independent of wavelength so the relative phase difference $\triangle\varphi$ will be proportional to the inverse of wavelength. This means that such a retarder will function as intended only at one specific wavelength. *Achromatic retarders* for use in the visible range may be assembled by combining typically two or three single retarders.

Retarders which generate a phase difference between the ordinary and extraordinary waves of $\pi/2$ are known as *quarter-wave plates*. This type of

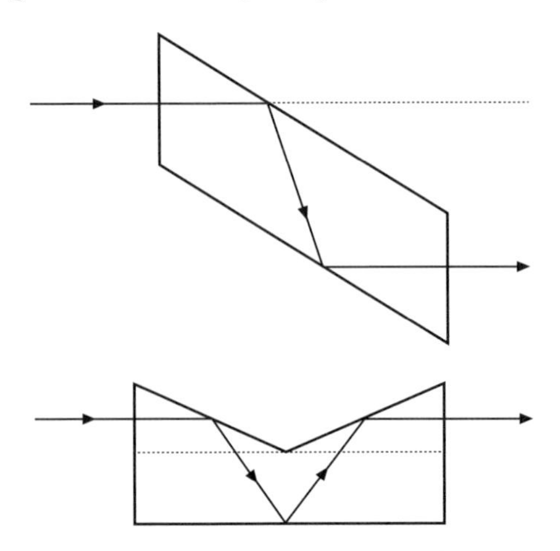

Fig. 9.21. Two quarter-wave achromatic Fresnel's rhombs

retarder generally serves to change linear polarization into elliptical as shown in Fig. 9.20. This process also works in reverse, i.e. elliptically polarized light may be converted into plane-polarized light. If linearly polarized light arrives at a quarter-wave plate at an angle of 45 degrees to either the fast or slow axis then the result will be circularly polarized light.

Correspondingly, a *half-wave plate* generates a phase shift of π. This causes a change in orientation of the plane of vibration of linearly polarized light. In the cases of circularly and elliptically polarized light the handedness of the polarization state changes. Furthermore, the main axis of an E-state ellipse assumes a new orientation.

A phase shift between the orthogonal components may also be brought about through total reflection. *Fresnel's rhombs* presented in Fig. 9.21 correspond to a quarter-wave plate. However, Fresnel's rhomb has one important advantage – it functions virtually as an achromatic quarter-wave plate but over a wide spectral range. Components such as these are generally made out of isotropic materials such as glass or fused silica. The Fresnel's rhomb in upper part of Fig. 9.21 effects a lateral shift between incoming and exiting waves. Reflection-based retarders may also be made in such a way that lateral shift does not occur (see lower part of Fig. 9.21). This kind of quarter-wave plate may also be designed to operate in the UV range and is particularly suitable for performing spectroscopic ellipsometry.

We have already noted above that one of the factors upon which retardance is dependent is the distance d travelled by the light within the retarder. By placing two wedge-shaped crystals of, for example, calcite adjacent to each other in the manner shown in Fig. 9.22, the polarized waves travel a distance in the system of $d_1 + d_2$. The desired retardance may be adjusted by altering the mutual position of the wedges. The phase difference may be calculated from the formula

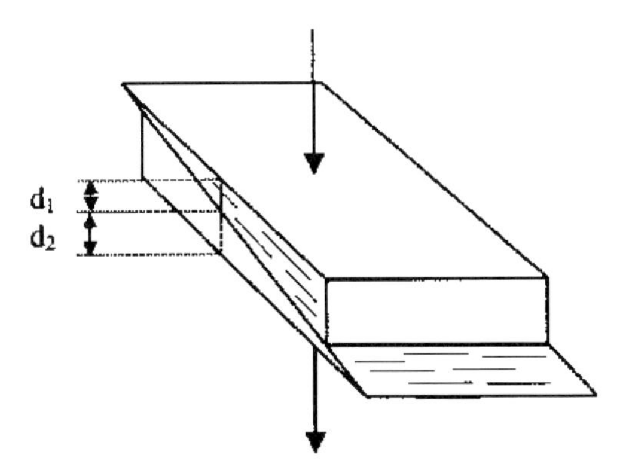

Fig. 9.22. A Babinet compensator. The desired phase difference is obtained by moving one wedge with respect to the other. The arrow represents the light path and lines on the wedge indicate the direction of the optic axis of the crystal

$$\triangle\varphi = \frac{2\pi}{\lambda}(d_1 - d_2)|\triangle n| \, . \tag{9.25}$$

The system presented in the diagram is known as the *Babinet compensator*. The optic axes of the narrow wedges are perpendicular to each other. Consequently, if $d_1 = d_2$ then each wedge will cancel the retardance brought about by the other and so $\triangle\varphi = 0$. The waves refract away from their original direction on encountering the oblique surface inside the compensator. This undesirable effect may be minimised by making the compensator as narrow as possible and, therefore, the wedge angle correspondingly small (in practice, it is a few degrees). Retardance changes linearly in the direction of length of the wedge. Being of some finite diameter, the incident light beam therefore experiences varying magnitudes of retardance along the wedge direction. The same phase difference may only be obtained from a narrow band in the direction perpendicular to that of the wedge direction.

Commercial compensators made from crystalline quartz such as the Babinet–Soleil compensator operate over a wide wavelength range, i.e. typically 250–3500 nm. Adjustment of retardance is accomplished mechanically using a micrometer screw.

Of course, mechanical and manual adjustment is inadequate if fast operation of the compensator is required. For such purposes electrically and magnetically controlled compensators have been developed. The active material used in electrically controlled devices may be either a solid or liquid. The devices most commonly used are the *Pockels cell* and the *Kerr cell*. These are both capable of rapid functioning. For example, a solid-state Pockels cell may perform a quarter-wave retardance shift in a few nanoseconds.

Depolarizers

Some applications require unpolarized, i.e. natural light. However, completely pure natural light is not readily available without the intervention of specific procedures or devices. As stated earlier many optical and mechanical components not intended to affect polarization nevertheless cause a partial polarization of light. In order to remove polarization we must use a device known as a *depolarizer*.

One method of depolarizing light is to take advantage of the multiple scattering of light. To this end we may use, for example, opal glass or integrating spheres. However, the applications of these components are limited because they completely destroy the geometry of the incoming light.

Since it may be difficult to use or even find genuine depolarizers we have to resort to the use of so-called *pseudo-depolarizers* in their place. These operate according to the principle that natural light may be considered as comprising the sum of all randomly occurring polarization forms. In a pseudo-depolarizer the form of polarization of the wave changes rapidly with respect to time, wavelength, location or some other non-essential quantity relating to the experimental arrangement. Since the output of such a system comprises the average of a large group of differing forms of polarization it may be reasonably considered to be unpolarized.

This idea is applied in e.g. the *Lyot-depolarizer* in which several randomly oriented retardation plates are placed in series. The mixing of the polarization achieved is a result of the wavelength spread and angular range.

9.4 Detectors

Detectors are devices which convert the useful information carried in electromagnetic waves into forms which can be more easily handled and interpreted. Earlier the most commonly used detectors were the human eye and the photographic plate. Nowadays, almost without exception, radiation is converted into an electrical signal, either a current or a voltage. Often this analogue signal is further transformed into digital form, allowing the information to be saved in a computer memory for further analysis and visual data study. An ideal detector is one which may be used over a wide wavelength range, which is highly sensitive and has a high signal-to-noise ratio, which reacts immediately to changes in the incoming signal and produces a linear response which is independent of wavelength. If in addition to all these features your detector is durable and sensibly priced then you indeed have excellent transducer.

Here we will examine some of the more important varieties of UV/visible range detectors. Our survey concentrates, however, on two particular types of quantum detector, the *photodiode* and the *photomultiplier tube (PMT)* as shown in Fig. 9.23. These two share the facility of being able to detect the intensity of light as well as intensity fluctuations almost in real time.

Fig. 9.23. A photograph of two photodiodes (plastic and metal packaged) and a photo multiplier tube (PMT)

Even today the average of light intensity is successfully and most simply represented by photographic film. This once highly popular form of detector may still be used in e.g. the measurement of emission spectra generated by ionised atoms. The integration and multichannel features of photographic film may also, of course, be reproduced by modern-day electronic detectors. Devices suitable for such purposes are array and matrix detectors made from semiconductors as well as the various types of integrators. Detector for special applications may consist of several elements and be fitted with appropriate filters as shown in Fig. 9.24.

Transducers used in optics may divided into two groups according to their principles of operation. These are *thermal detectors* and *quantum detectors.* As their name suggests, thermal detectors react to heat. The consequences of this reaction are detectable in terms of changes in electrical resistance, electrical polarization or electromotive force. Although this kind of detector is mostly used in the IR range it also has some applications in the visible range, e.g. laser power measurements. By contrast, photon detectors (also known as *photoelectric detectors*) react in a quantum manner to incoming radiation, i.e. to photon flux. In the ideal case the detector reacts to the arrival of a photon by generating a single *response unit.* This response unit may take the form of an electron (commonly called a photoelectron) freed by the photon in a photomultiplier tube or, alternatively, an electron-hole pair generated in a semiconductor. In all cases the rearrangement of the charges produces either an electric current or a voltage. It should be noted that, in contradistinction to the case of the thermal detector, photons possessing differing energy, i.e. of differing wavelengths, in principle produce identical

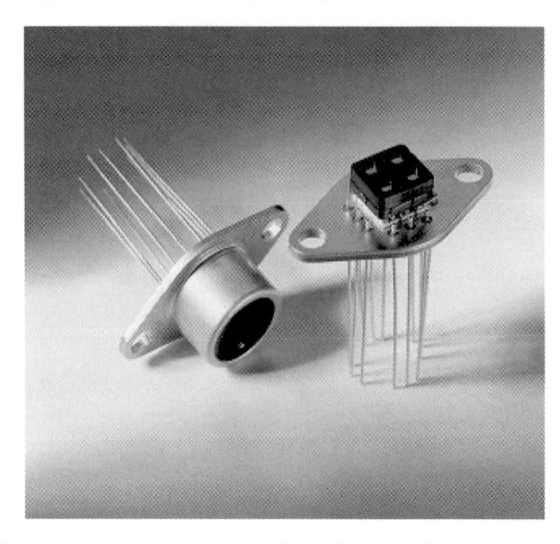

Fig. 9.24. A four-channel detector module for the IR range (Photo courtesy of VTT Electronics)

units of response. Also, quantum detectors are, in general, more sensitive than thermal detectors although their wavelength response is less even.

9.4.1 Performance Parameters

The performance of detectors may be evaluated using a wide range of parameters. These also help us to choose the right detector for a given requirement. In general, these values depend on wavelength and temperature as well, in some cases, as the surface area of the detector, the modulation frequency and bias voltage. Let us assume that the detector is a current source, which is, indeed, most commonly the case.

Responsivity (or *photosensitivity*) S is usually defined as the output/input ratio of the detector. In the case of light measurement the most obvious unit of responsivity is either the ampere per watt (A/W) or ampere per lumen (A/lm). Responsivity is wavelength-dependent, especially in the case of quantum detectors. Responsivity may be expressed either in absolute terms or as a relative value, i.e. normalised with respect to the maximum responsivity located at some wavelength.

Every photon arriving at an ideal quantum detector produces a response unit. In practice, however, this does not happen as devices fail to respond to all incoming photons. The ability of a detector to utilise incoming photons is described by the concept of *quantum efficiency* QE. This is the fraction of a photoelectron created per incident photon. This may also be expressed in terms of responsivity S (A/W) using

$$QE = 1240 \frac{S}{\lambda} \cdot 100\% \,. \tag{9.26}$$

Here wavelength λ is measured in nanometres.

The photocurrent is dependent not only on the intensity of light arriving at the detector but also on the wavelength of the light. This wavelength dependence is described by the term *spectral response* and may be presented in terms of either responsivity or quantum efficiency.

Thermal detectors in particular are unable to react to rapid fluctuations in radiation intensity and their operation inevitably incorporates an element of delay. Reaction capability is described in terms of *response time* or *rise time* τ. This is the time taken for the output signal to rise from one previously determined level to another following a stepped light input. By convention, this is calculated as the time taken for the signal to rise from 10–90% of the total leap. Depending on the type of detector the rise time may vary from a matter of femtoseconds to seconds.

If the arrival of light at the detector is suddenly interrupted then the signal begins to fall off towards zero. The *time constant* is defined as the time taken for the signal to fall to e^{-1} (i.e. to about 37%) of its original level. The time constant and rise time generally differ in value.

Similarly, the *cut-off frequency* f_c describes the time response of a photodiode. Sine wave modulated light is used to determine this value. The cut-off frequency is the frequency at which the signal is damped 3 dB with respect to the reference level. The cut-off frequency is affected by wavelength, the resistance of the load and the frequency defining the reference level, typical values for these being $\lambda = 830\,\mathrm{nm}$, $R = 50\,\mathrm{ohms}$ and $f_{\mathrm{ref}} = 100\,\mathrm{kHz}$. Clearly, rise time is inversely proportion to the cut-off frequency.

The *noise equivalent power* (NEP) of a detector describes the lowest intensity of light which the detector is capable of measuring, i.e. the optical power for which the signal-to-noise ratio of the system is equal to one. Since NEP describes power it is measured in units of watts (W). We will discover in Chapter 10 a connection between the magnitude of noise and the frequency bandwidth. For this reason it is helpful to use the *normalised* NEP value for unit bandwidth. This is measured in units of $(\mathrm{W}/\sqrt{Hz})$ and is often denoted as NEP*.

Detectivity D is defined as the inverse of NEP. Both of these are affected by the surface area A_D of the detector and the bandwidth $\triangle f$. In order to facilitate comparison between detectors of different sizes a useful term known as the *area normalised detectivity* D^* is employed. This is expressed by

$$D^* = D(A_D \triangle f)^{1/2} = \frac{1}{NEP}(A_D \triangle f)^{1/2} . \tag{9.27}$$

The range of light power which produces a linear response in a detector is known as the *dynamic range*. The upper limit of the linear range is generally determined by the electrical characteristics of the detector and the external circuit. Correspondingly, the lower limit is indicated by the NEP value.

9.4.2 Thermal Detectors

The functioning of thermal detectors is based on the dissipative absorption of radiation. The heat generated alters the physical and electrical properties of the active material in the detector. These changes may be detected by connecting an electrical circuit to the system. The output of the detector is proportional to the amount of energy absorbed per unit time.

Considerable demands are made on thermal detectors as the temperature changes which occur are often small. Thermal detectors are passive devices and do not therefore require biasing. Moreover, they are stabile. Thus, for example, the thermopile may be used for calibration purposes.

When two dissimilar metals are connected together a small electric potential forms across the junction. The magnitude of this potential is of the order of millivolts and is dependent on the temperature of the metals at the junction point. It is, therefore, possible to use the system for measuring temperature. Voltage fluctuations are of the order of $50\,\mu V/°C$. For application in the IR range thin wires made from bismuth and antimony are used. In order to increase absorption the junction point between the wires is coated with black paint. Often a reference junction shielded from the radiation to be measured is fitted inside the detector. This allows measurement of the temperature difference between the active and reference junctions. Transducers operating according to this principle are known as *thermocouples.*

The voltage signal generated by a single thermocouple is weak. It may be augmented by connecting several thermocouples in series. A *thermopile* typically contains tens of thermocouples. It is capable of detecting optical power having normalised NEP value of 10^{-9} $W/\sqrt{Hz}$ over a wide spectral range. The usefulness of the detector lies in the simplicity of both its construction and operation principle as well as in its robustness.

It is well known that an electric field will cause *electric polarization* in a dielectric material. Usually this polarization disappears rapidly when the external field is removed. However, some dielectric material possess remarkable electro-thermal properties, namely, they retain their polarization after the external field has been removed. This phenomenon is of great application since polarization in these so-called *pyroelectric materials* is highly dependent on temperature.

A change in the electrical polarization causes opposite charges to move to opposite surfaces of the element. When electrodes including electric wires are attached to the surfaces the system functions as a capacitor. Since the element responds only to temperature changes so the incoming light must be either modulated or in pulse mode. Reaction to the changes is relatively fast – pyroelectric detectors are also used in FTIR spectrophotometers, power meters and laser pulse detectors.

Pyroelectric detectors are more sensitive than thermopiles and are able to disregard steady background radiation. It is important to note that as temperature increases net polarization decreases and, indeed, disappears totally

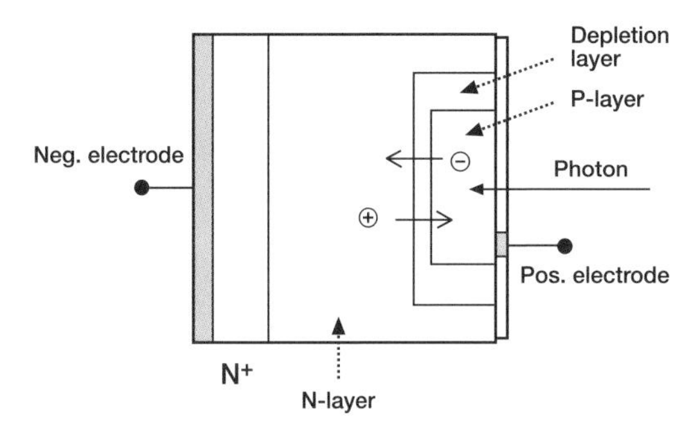

Fig. 9.25. A cross-section of a PNN^+ photodiode

on reaching the *Curie point.* One important pyroelectric material, triglycine sulphate (TGS) has a Curie point as low as 49°C. Other materials used include triglycerine selenate (TGSe), lithium tantalate ($LiTaO_3$), lead zirconate (PZT), strontium barium niobate (SBN) and polyvinyl fluoride (PVF).

9.4.3 Photon Devices

The active areas in quantum detectors also absorb incoming light. As a result electrons either move up to conduction bands (a process known as *photoconduction*) or are completely ejected from the active material. In both cases an electric current is generated which is proportional to the number of incident photons. This type of detector is used mainly in the ultraviolet, visible and NIR ranges.

Photodiodes

A *photodiode* is a semiconductor device which generates an electric current or voltage when light enters its active material. Photodiodes are typically used for measuring light intensity and determining the presence, location and colour of light. The extremely wide application of these components are a result of their excellent properties. These include linearity of electrical response and light intensity over a broad range, low noise, and operability over a wide range of the spectrum. Furthermore, they are mechanically robust, light, small in size and durable. Photodiode types include PNN^+, PIN, APD and Schottky.

Let us now consider the operation of the PNN^+ as shown in Fig. 9.25. Light quanta arriving at the component excite the electrons in the valence band of the semiconductor material. Electrons enter the conduction band providing that the energy they receive is greater than the required transfer energy or band gap energy E_g. On entering the conduction band each electron leaves behind a hole in the valence band. This process occurs in all

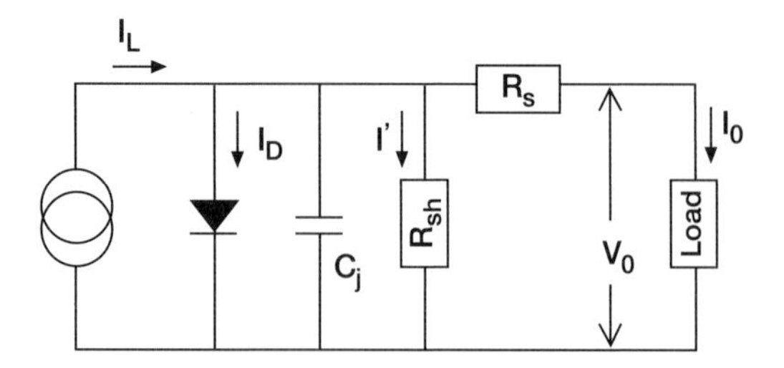

Fig. 9.26. An electric equivalent circuit of a photodiode

semiconductors in the p-layer, i.e. active area, in the n-layer and in the depletion layer. The number of electron-hole pairs generated is proportional to the number of incident light quanta.

An electric field transfers the electrons in the depletion layer to the n-layer and holes, correspondingly, to the p-layer. Those electrons generated in the n-layer remain in the conduction band as do those which arrive from the p-layer. Correspondingly, the holes generated remain in the valence band as they move through the depletion layer to the p-layer. Thus the charges are now ready to generate a current in the external conductors connected to the system.

The electrical functioning of a photodiode may be described by the *equivalent circuit* presented in Fig. 9.26. The output current I_0, that is, the current passing through the external load, may be calculated using

$$I_0 = I_L - I_D - I' = I_L - I_S \left[\exp\left(\frac{eV_D}{kT} \right) - 1 \right] - I' \, . \tag{9.28}$$

The terms in the equation are as follows: I_L is the current generated by the incident light, I_D the current passing through the diode, C_j the junction capacitance, R_{sh} and I' the shunt resistance and shunt current, R_s the series resistance, V_o the output voltage, V_D the potential across the diode, I_S the photodiode reverse saturation current, e the elementary charge, k Boltzmann's constant and T the absolute temperature of the diode.

Photodiodes may be made to operate very rapidly. The time response is described by the rise time and cut-off frequency. These parameters depend on, among other things, the wavelength employed, the resistance connected to the diode and the area of its active surface. One important factor affecting the time response of photodiodes is the terminal capacitance C_t. This is the sum of the junction and package capacitancies of the diode. The smaller the capacitance the faster the operation of the diode and *vice versa*. In high-speed applications such as optical communications and high-speed photometry APD and PIN diodes are used. These may operate at even gigahertz frequencies.

The transfer of electrons and holes is more rapid inside the depletion layer than outside it. Therefore, the best time response and strongest current are

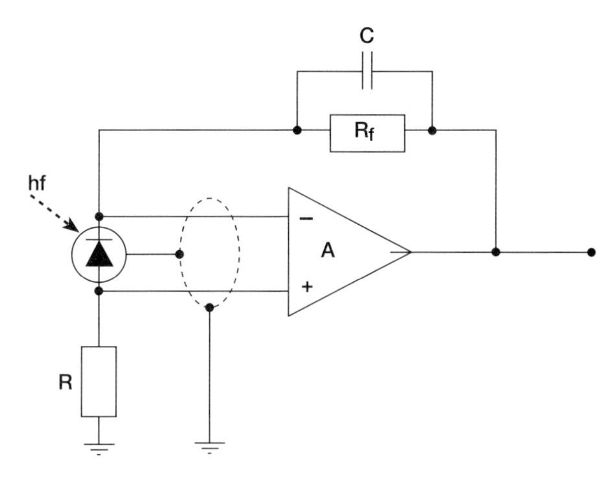

Fig. 9.27. A photodiode amplifier circuit for high-sensitivity detection (for example $C = 5$ pF, $R = R_f = 10^9$ Ohms, A = OPA128LM and photodiode HP 5082–4204 would yield the bandwidth of DC–30 Hz and the responsivity of 10^9 V/W [208])

achieved when photon absorption occurs inside the depletion layer. Fortunately, the thickness of this layer may be adjusted by the *voltage biasing* of the diode. In the *photovoltaic mode*, i.e. when no voltage biasing is applied, the depletion layer is very thin. If a positive voltage is applied over the p-n-boundary then the system is forward biased while a negative voltage results in reverse bias. In the latter case the photodiode is in the *photoconductive mode*. This mode achieves an improved time response although at the expense of increased noise.

When fitted with windows which transmit UV light photodiodes may operate over a spectral range which comfortably includes the UV and visible regions, i.e. 190–1100 nm. The cut-off wavelengths are determined by the properties of the photodiode material. The upper limit is dictated by the band gap energy E_g. For silicon this has a value of 1.12 eV and for GaAsP 1.8 eV. These energies correspond to wavelengths of 1100 nm and 700 nm, respectively. From this it follows that electromagnetic radiation of wavelength greater than 1100 nm is not able to produce electron-hole pairs in silicon semiconductors. As the wavelength shortens absorption occurs increasingly close to the surface of the semiconductor in the diffusion layer, thereby impairing the efficiency of the process. The lower cut-off frequency of standard detectors lies at around 320 nm while those designed for operation in the UV range extend as far as 190 nm.

By applying an appropriate electric circuit the operation of the photodiode may be optimised for the application in question. If a linear and a high-sensitivity response is desired we may use the circuit presented in Fig. 9.27. The linear range may extend over nine decades of light power, i.e. from 10^{-12} to 10^{-2} W.

Commercially available photodiodes come in a range of shapes and sizes. The dimensions of their active areas typically vary from a matter of millimetres to a few centimetres. Light diodes are excellent detectors providing that light intensity is sufficient. However, lower intensities cause problems. Let us consider a PIN diode being bombarded by a thousand photons per second. This is still detectable by the human eye but the current it generates in a PIN diode is only approximately 0.4 fA. Such as small signal is inevitably lost amidst the background noise.

Photomultiplier Tubes (PTM)

Photomultiplier tubes are the most sensitive of all detectors over the UV/visible range. At their best these ingeniously functioning tubes are able to detect optical powers of below 10^{-15} W [209]. Furthermore, the response of the detector to incoming radiation change is rapid. Owing to their photosensitivity they may be used not only for light intensity measurements but even for photon counting.

PMTs comprise generally of a photocathode, focusing electrodes, an electron multiplier and an anode. These are all contained within a glass vacuum tube. In Fig. 9.28 we present the basic features of two PMT constructions. In photoemissive detectors such as the photomultiplier tube light interacts with the electrons of the detector material, i.e. the photocathode. If the energy hf of a photon arriving at the detector is greater than the work function Φ_0 of the photocathode, i.e. $hf > \Phi_0$, then an electron will be released from the atom. In addition, the energy difference $hf - \Phi_0$ imparts kinetic energy on the electron and so removes it from the photocathode. This phenomenon is known as the *photoelectric effect.*

Free electrons produce a weak cathode-photocurrent. This current is amplified in the electron multiplier component by factors ranging typically from 10^5 to 10^7. On entering the multiplier the electrons are accelerated by an electric field and are allowed to collide with the intermediate anodes (known as dynodes). Each colliding electron possesses considerable kinetic energy and is able to release a number of new electrons from the intermediate anode. These new electrons are now ready to travel on to successive dynodes which are placed in order of increasing electrical potential. After about ten such multiplications a large number of electrons arrive at the anode in the form of a current along the tube.

The amplification process is rapid and produces very little noise. Rise times for such tubes are typically measured in nanoseconds. The responsivity of an average photomultiplier tube is of the order of one ampere per microwatt. Under normal conditions the current travelling along the tube is relatively small, i.e. a few microamps while the maximum current should be fixed at less than a milliamp.

A tube containing M dynodes has a current amplification coefficient G of

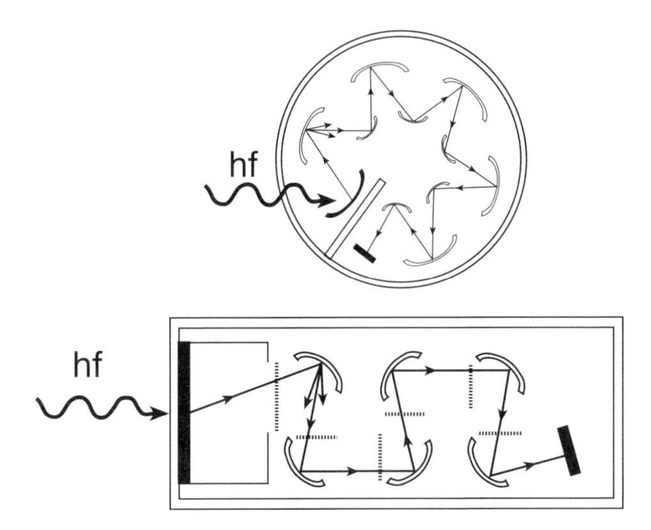

Fig. 9.28. Two PMT configurations; end-on box and grid (lower) and side-on circular cage focused (upper)

$$G = \delta^M = KV^{\eta M} , \tag{9.29}$$

where δ is the average number of secondary electrons, K is a constant, η is a constant dependent on the material and geometrical structure of the dynode (about 0.7–0.8) and V is the potential between the anode and cathode. Secondary emission depends on the dynode material and the potential applied. For example, with an interdynode voltage of 200 V a Be–Cu dynode will produce an average of six new electrons for each incident electron. According to (9.29) the eight dynodes in question would produce a current amplification of about $1.7 \cdot 10^6$. The voltage used in PMTs varies between 400 and 1500 V. The voltage between the cathode and first dynode must be large enough for the tube to function linearly.

The multiplication of electrons should take place in a controlled and reproducible manner. This requires that the trajectories of electrons remain unchanged. The flow of electrons may be disturbed by external magnetic fields. A stable power source is also essential for successful operation. Photomultiplier tubes are extremely sensitive. They should not be handled and certainly not powered up in bright daylight.

A weak current may be observed in the tube even when no light is entering it. This so-called *dark current* is generated mainly from electrons released from the cathode purely by virtue of their thermal energy. As a consequence we may observe at room temperature a few tens of pulses per second for each square centimetre of the photocathode surface. This may cause problems in photon counting or in low-light-level detection. On the other hand two picowatts incident power causes some 10^6 pulses per second! If necessary, however, dark current may be reduced by lowering the temperature of the detector.

Dark current may also be generated by the ionisation of the residual gas in the tube, scintillation in the surrounding glass envelope or field emissions. It may also arise a leakage current as a result of imperfect insulation.

When studied externally photomultiplier tubes may be divided into two groups on the basis of their construction. These are known as *side-on* measuring and *end-on* measuring tubes. The first of these generally contains a reflecting photocathode while in end-on measuring tubes the photocathode is semi-transmitting. The area of the cathode surface sensitive to light may be large, i.e. tens or even hundreds of square centimetres. The dynodes may be placed in a variety of spacial configurations with respect to one another. Figure 9.28 shows the *end-on box and grid* and *side-on circular cage focused* dynode geometries.

PMTs operate over the entire UV/visible range. The upper limit of the useful wavelength range may be adjusted through the choice of appropriate photocathode material. In general these are alkali metals with low work functions. Materials used include Ag-O-Cs, Ga-As, In-Ga-As, Sb-Cs, Na-K-Sb and Na-K-Sb-Cs. The quantum efficiency of the best photomultiplier tubes exceeds 25%. The lower limit of the wavelength range is determined by the properties of the window material. Favoured window materials include borosilicate, UV glass and magnesium fluoride.

Further information of detectors may be found e.g. in [210, 211].

10 Understanding Your Signal

Manmade devices intended for performing measurements of even the slightest complexity operate to a greater or lesser extent non-ideally. The final result includes an error component, the origin of which may be traced to inadequate calibration, noise, stray light, incorrect data processing or user error. In addition, careless sampling and sample processing cause problems. In this chapter we will look at indicators of measuring device performance, sources of error and methods for eliminating noise.

10.1 Calibration Curve

The relationship between the quantity to be measuredand and output of a device is given by the experimentally obtained *calibration curve* (Fig. 10.1). Thus, for example, the calibration curve may indicate the mutual dependency of temperature and the voltage signal produced by the measurement device or it may give transmittance values using the current passing through a photomultiplier tube. The calibration curve may be determined by changing the measurable variable from minimum to maximum and back again. This operation should be performed several times. In favourable cases the calibration curve will take the form of a straight line. On other occasions it may include, amongst other things, defects in form or hysteresis. Nonlinear calibration curves may nevertheless be rendered linear using various approximations such as line segments, polynome or function approximations, bilinear expressions or simply tabular corrections [212].

The degree of nonlinearity of a measurement device may be shown in a number of ways depending on how the reference line is placed with respect to the calibration curve. The three most common cases are presented in Fig. 10.2. They are

- Independent nonlinearity
- Zero-based nonlinearity
- Terminal-based nonlinearity

In the first of these cases the reference line is chosen such that the maximum deviation of the calibration curve with respect to the reference line is as small as possible. In the second case the reference line is fixed to the initial point

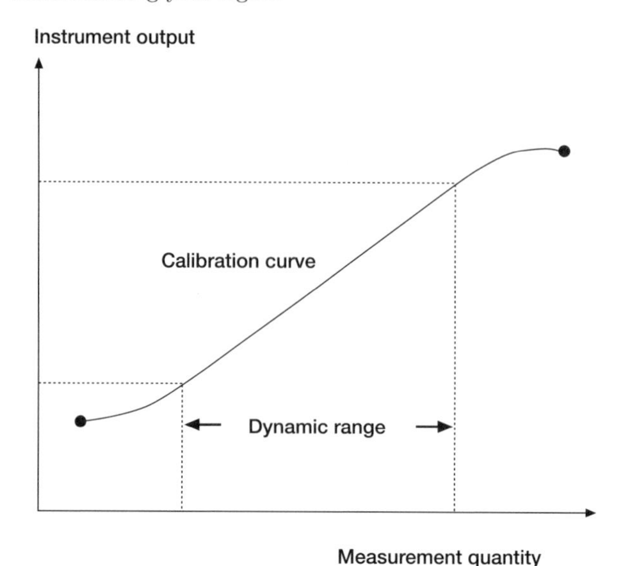

Fig. 10.1. The output of an instrument and the measurement quantity obtained are linked together via the calibration curve. The linear part of the curve is known as the dynamic range

of the calibration curve. At the same time, maximum deviation between the two lines is kept to a minimum. In the case of terminally-based nonlinearity the reference line passes through both the initial point and terminal point of the calibration curve. In this way nonlinearity is obtained from the maximum deviation. Also of importance in the placing of the reference line is *zero error*. This is defined as the deviation at the initial point of the measurement range. From such a definition it follows that the zero error disappears in cases of zero-based nonlinearity and terminal-based nonlinearity.

In addition to the presence or absence of linearity the calibration curve is also affected by numerous other factors characteristic of the measurement process (Fig. 10.3). The term *dead band* is used to indicate the extent to which the measured quantity may vary from the chosen value without the output value changing. Correspondingly, the *resolution* of an amplitude-discrete measurement device refers to the range of the measurable variable over which output remains constant.

A decrease or increase of the measurable variable, in other words, its alteration in either direction, may produce two different calibration curves which meet at either end of the measurement range. This effect is known as *hysteresis*. Its magnitude describes the maximum separation between the two curves. Also relating to the calibration curve is the concept of *repeatability* or *precision*. Indeed, this often comprises the maximum difference between outputs for a given value of the measurable variable. In such cases observations must be taken in constant conditions. Likewise, in determining precision no

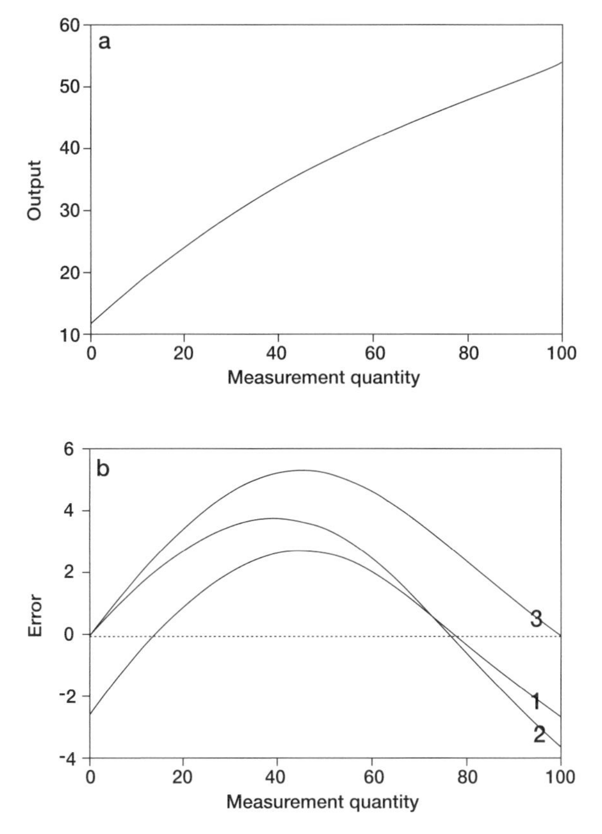

Fig. 10.2. (**a**) A nonlinear calibration curve for an imaginary system. The degree of nonlinearity depends on how the reference line is oriented respect to the calibration curve. (**b**) The error curves obtained from (1) independent, (2) zero-based and (3) terminal-based nonlinearity

account is taken of hysteresis, i.e. it is determined using the same branch of the calibration curve.

10.2 Indicators of Performance

A thorough examination of the calibration curve reveals much useful information about the measurement device to its user. Likewise, the so-called *figures of merit* describe the technical performance of the device. They may be used to determine, for example, whether or not it is suitable for the study and measurement of a particular material. Of course, the choice of measurement device also involves consideration of other factors such as the speed of measurement or analysis, ease and convenience of performance, cost of tests and requirements of sample handling.

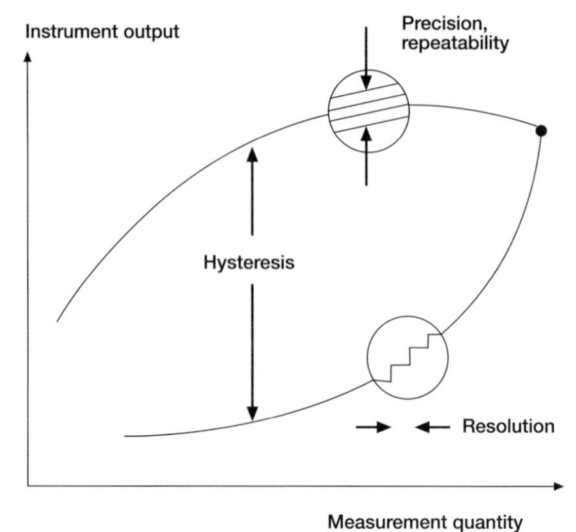

Fig. 10.3. A closer look of the calibration curve reveals non-desirable features characterised by, e.g. hysteresis, precision and resolution

Both spectrophotometry and reflectometry are widely exploited in analytical chemistry. Indeed, one of the main applications of these techniques is in the determination of concentrations. According to Skoog, Holler and Niemann [187], the most important figures of merit in instrumental analysis are precision, bias, sensitivity, detection limit, concentration range and selectivity. Below we examine each of these from the perspective of measurement of concentration.

10.2.1 Precision

The correctness of measurement results or the reliability of data is generally described by two terms, *precision* and *accuracy*. As we have already stated, precision describes the degree of deviation exhibited by the measurement output. In practice this means that measurements are repeated in an identical fashion several times in succession. Following this the deviation of the obtained data is described using statistical parameters such as standard deviation, variance and coefficient of variation.

Accuracy describes how close the experimental result is to the true or correct value. This term may be used either as a general term or in referring to some specific accuracy class. In detailed presentations it is better to use the term error. In quantitative terms accuracy is problematic because the determination of the correct result is impossible except in certain special cases such as object counting. In practice the correct result may be taken to be that obtainable using generally approved standard procedures or devices

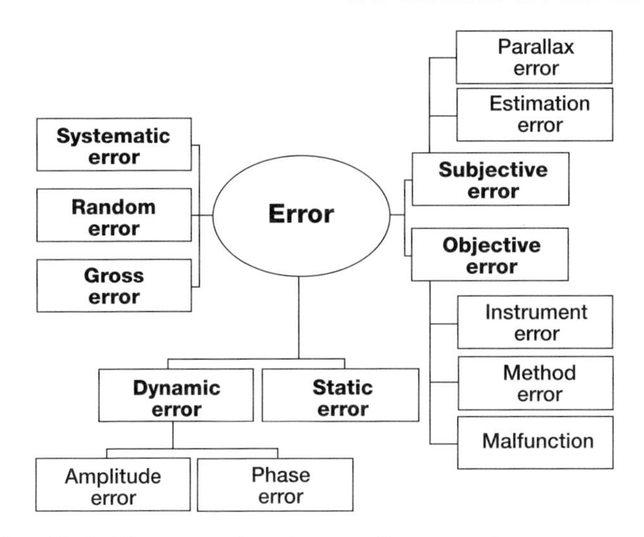

Fig. 10.4. Three ways to categorise the error of a measurement

which are known to be more accurate and reliable than the device being used. Accuracy or error is expressed either as absolute error E_{abs} or relative error E_{rel}

$$
\begin{aligned}
E_{\mathrm{abs}} &= \overline{x} - x_t \\
E_{\mathrm{rel}} &= \frac{\overline{x} - x_t}{x_t} \cdot 100\% \,.
\end{aligned}
\tag{10.1}
$$

In the above formulae $\overline{x}$ represents the average value of the experimental results and x_t the true result obtained using a reference method.

Error may be classified according to the origin of the error or by some other reference method (see Fig. 10.4). Inaccuracy indicates the systematic error which should be corrected by whichever means available. Following correction procedures there remains uncertainty which is mostly made up of random error.

Typical systematic errors are connected to device errors such as wear of device components, electronic drift, changes in the operating voltage and the effects of temperature on the detector. The user may misread the scale indication. In addition, the performer of the test always has some kind of expectation or hope concerning the final result and this may lead the person to round off the measurement values obtained in a particular direction. Errors of method usually arise from the nonlinear behaviour of the sample material. The presence of such errors may be identified by validation analysis. This involves the use of standard samples possessing characteristics similar to those of the original samples.

In maintaining accuracy of measurement it is important to observe changes occurring over long periods of time. These are collectively known

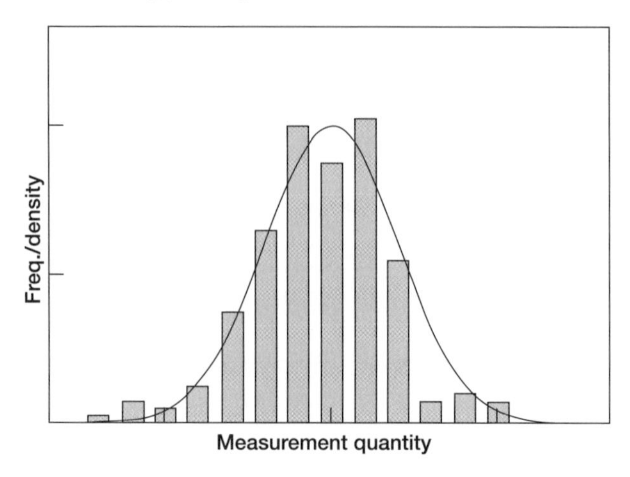

Fig. 10.5. The distribution of several consecutive measurements (dark bars) resembles a normal error curve (solid line). This is an indication of random error

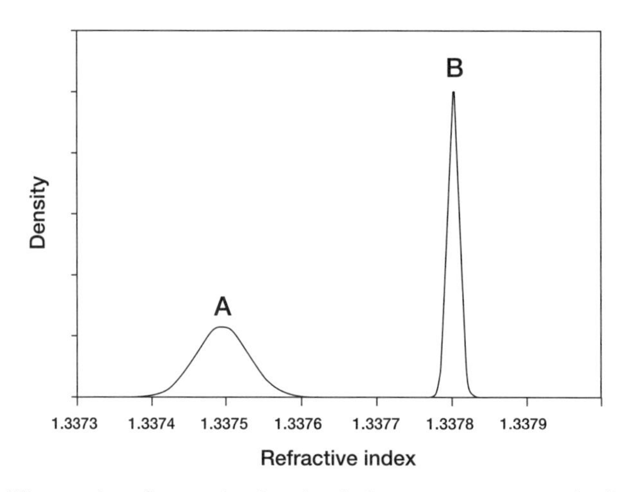

Fig. 10.6. The results of several refractive index measurements of a liquid sample obtained using two apparatuses. The random error of device B is smaller to that of device A

as *drift* or *instability*. *Reproducibility* is the largest observed variation in output over a specific time period and in constant operating conditions. Reproducibility is affected by precision, hysteresis and instability.

When output is used to control some process or to measure rapidly changing values the dynamics of the measurement system must be known. Dynamics and associated errors are described using the concepts of step response and frequency response.

Individual measurement results x_i may be used to build up a histogram (frequency against value). As shown in Fig. 10.5, the distribution described by the histogram very often resembles a Gaussian curve, i.e. a normal error curve. In such cases the measurement data includes random error. The curves in Fig. 10.6 are results of trials performed using two refractometers. It should be noticed that the refractive indices given by device B are much more densely centred around their average than are those of device A. If the data does not contain any other type of error than random error then the peak of this distribution curve represents the true result. It follows that the effect of random error may be significantly reduced by increasing the number of measurements and calculating their average value. In practice between twenty and thirty measurements are sufficient to yield a good result.

We have already mentioned random and systematic errors E_r and E_s. We may define yet a third type of error, namely, *gross error* E_g. This includes the type of random errors caused chiefly by the laziness or carelessness of the user during the measurement process. Such measurement results often differ greatly from the others and so may be removed from the data with impunity. Absolute error may thus expressed as follows:

$$E_{abs} = E_r + E_s + E_g \ . \tag{10.2}$$

10.2.2 Bias

The mechanism by which *bias* arises in measurement techniques may in principle be determined. It therefore belongs to the class of systematic errors. This means that for identical measurements the error will have a specific value and sign. If the population mean of the sample is μ and x_t is its true value then the *bias* is given by

$$bias = \mu - x_t \ . \tag{10.3}$$

Suppose that the average of the curve on the right hand side of Fig. 10.6, (device B) corresponds to the true, error-free refractive index of a liquid under investigation. We may then obtain the bias for device A from the difference between the peaks of the two curves. In this particular case it is approximately 0.0003 units of refractive index.

10.2.3 Sensitivity

Sensitivity expresses quantitatively how small a measurable quantity (e.g. differences in concentration) the device or method is capable of detecting. If we examine the calibration curve in Fig. 10.3 we notice that the sensitivity of the device is limited by its precision. Also the steepness of the calibration curve, i.e. its slope, exerts an effect on sensitivity – a steeper curve produces greater sensitivity and *vice versa*. The slope of the curve m is also known as the *calibration sensitivity*. If the calibration curve is linear the measurement signal S (in this case S depends on concentration) may be written in the form

$$S = mc + S_{bl} . \tag{10.4}$$

In this formula c is the concentration of the analyte and S_{bl} the signal of a blank sample. *Analytical sensitivity* γ also takes account of the precision of the device. If s is the standard deviation of the measurements then analytical sensitivity may be determined as follows

$$\gamma = m/s . \tag{10.5}$$

Even though γ is dependent on concentration it has the advantage of being almost completely unaffected by changes in the amplification of the device and is also independent of the units used to measure the signal.

10.2.4 Measurement Range, Concentration Range

The lower limit of the range of the measurable quantity, the *detection limit*, c_m, is the minimum concentration which may be detected at a given confidence level. A quantity may be detected with a good degree of certainty (confidence level of 95%) when the amplitude of the measurement signal is at least three times larger than the standard deviation of a blank sample [187, 213]. From this it follows that the detection limit is determined by a ratio of three times the blank sample signal to the slope of the calibration curve.

The *limit of quantitation* (*LOQ*) is the lowest concentration at which quantitative measurement may be performed. Roughly speaking, the limit of quantitation may be taken to be ten times the standard deviation of a blank sample, i.e. $LOQ = 10s_{bl}$. Measurement techniques aim to ensure that the linearity between the measured signal and the measurable variable holds good over a wide range (at least two decades). The range beginning at the limit of quantitation and terminating at the upper limit of linear response, is known as the *dynamic range*.

10.2.5 Selectivity

Often the sample to be studied is a complex containing numerous elements and their compounds. If it is our intention to determine one particular component in such a sample then it is likely that the other components present will hinder the final result of the measurement. The term *selectivity* describes the ability of the measurement technique to disregard the components not of interest to the study in question. This matter may be illustrated by comparing the slopes of the calibration curves of two components. Let A be the component of interest and B the undesirable component. The ratio of the two slopes is known as the *selectivity coefficient* $k = m_B/m_A$. In the ideal case k would be zero. The selectivity coefficient may be either positive or negative. In the latter case the disturbing component reduces the signal produced by the device.

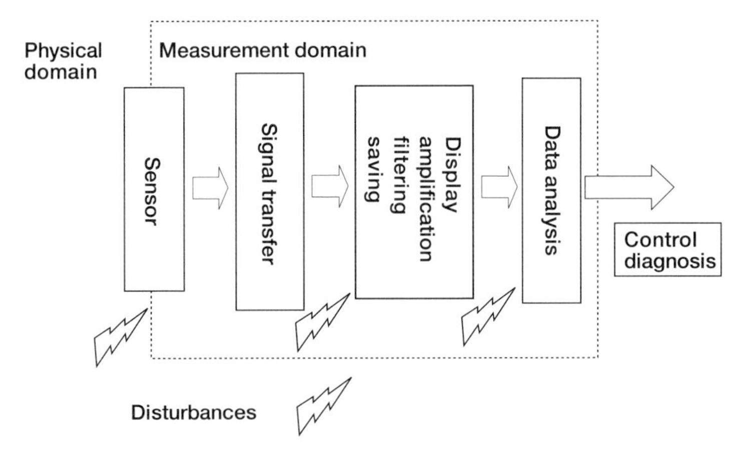

Fig. 10.7. A measurement system characterised according to its main functions. The signal is likely interfered by disturbances

10.3 Sources of Noise and Error

Figure 10.7 presents a general scheme for a measurement system. The signal is generated when the sensor locates a measurable variable in the physical domain. The signal is transferred to the measurement domain of the system where it is processed in a number of ways. The final result involves the system supplying the user with the required information or sending it on to, for example, a regulating or control device. It is important to notice that disturbances are associated with operations in both physical and the measurement domains of the system.

The measurement signal may in principle be divided into two components, the information required and the undesirable noise. Clearly the latter has a tendency to distort the final result. Noise may be characterised according to its amplitude and frequency distribution or alternatively by the physical mechanism which produces it. The contribution of noise is described by the *signal-to-noise ratio* S/N. A measure for noise may be taken from the standard deviation of numerous measurements of a DC signal. Thus

$$S/N = \frac{\overline{x}}{s} \,. \tag{10.6}$$

Here $\overline{x}$ is the average signal and s its standard deviation. In electronics the ratio of two quantities (e.g. voltage) is usually presented in units of decibels. If the root-mean-square voltages of noise and the desired signal are V_n and V_s we may write the S/N ratio in the form [214]

$$S/N = 10 \log_{10} \left(\frac{V_s^2}{V_n^2} \right) \mathrm{dB} \,. \tag{10.7}$$

Noise may be frequency dependent. One such example is flicker noise, the magnitude of which is inversely proportional to frequency. This kind of noise is termed *pink noise.* If there is no frequency dependence we speak of *white noise* (note the analogy to white light which contains all frequencies of the visible spectrum).

Modern-day spectrophotometers contain a great deal of electronics. The purpose of this, amongst other things, is to reinforce and filter the signals, to control various operations and to modify and present an analogue signal in digital form. Unfortunately, each of these operations adds an undesirable voltage or current component to the true signal. Noise attaches itself to the signal at a various places and in various ways. Because noise generated in this manner is often complex its characterisation and analysis can be problematic. We will now consider the most common forms and sources of noise affecting measurement. These include thermal noise, shot noise, flicker noise, environment noise, stray light and chemical noise.

10.3.1 Thermal Noise or Johnson Noise

Heat causes the random motion of electrons and other charge carriers and thereby voltage fluctuation in resistive components. Consequently, thermal energy causes a small noise voltage i.e. *thermal noise* to arise between e.g. the terminals of an isolated resistor.

From a thermodynamic perspective it may be shown that in an open circuit the root-mean-squared noise voltage V_n at temperature T (K) is given by

$$V_n = \sqrt{4kTR\triangle f}\ . \tag{10.8}$$

In this equation k is the Boltzman constant, R the resistance of the component and $\triangle f$ the frequency band of the signal. Although the amplitude of thermal noise voltage is random and therefore unpredictable it nevertheless exhibits a normal distribution (Fig. 10.8). The probability $p(V)d(V)$ that at some arbitrary moment in time the noise voltage will lie within the voltage interval $[V, V + dV]$ may be calculated using the equation [214]

$$p(V, V + dV) = \frac{1}{V_n\sqrt{2\pi}} \exp\left[-\left(\frac{V}{2V_n}\right)^2\right] dV\ . \tag{10.9}$$

Thermal noise disappears only at the ideal temperature of absolute zero. Otherwise it is always present. It may be reduced by lowering the resistance of the components. Here, however, it is important to note that thermal noise is independent of the physical composition of the resistor. Furthermore thermal noise is present even when no current is passing through the component.

Thermal noise may also be reduced by narrowing the frequency band. However, a narrower frequency band causes the device to react more slowly to changes in the signal. This follows from the fact that the quantity which

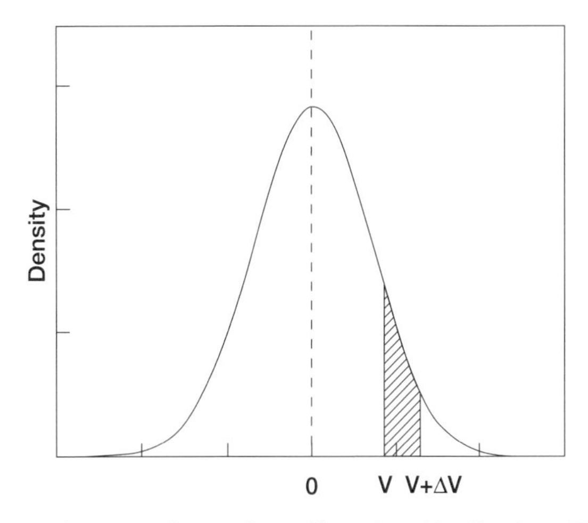

Fig. 10.8. Thermal noise voltage obeys Gaussian distribution. The shaded area indicates the probability of an instantaneous thermal noise voltage lying between V and $V + dV$

describes the response speed of the device, rise time t_r, is inversely proportional to the frequency band, i.e. $t_r \propto 1/\triangle f$. Thus more time is required for performing reliable measurements. If, for example, we wish to reduce thermal noise by a factor of ten then we need to reduce the frequency band by a factor of 10^2, e.g. from 1000 Hz to 10 Hz. Although thermal noise is clearly dependent on the frequency band it is, nevertheless, white noise. In other words, it is independent of frequency itself.

Another commonly used method of reducing thermal noise is the reduction of the temperature of the detector. This may be achieved using e.g. liquid nitrogen. At the temperature of liquid nitrogen, i.e. approximately 77 K, the level of noise falls to a half of its room temperature value.

10.3.2 Shot Noise

Shot noise is based on the quantum nature of the quantities to be measured. An observable example is provided by the arrival of individual light quanta at a photodiode. Each quantum causes the generation of an electron-hole pair and finally the transfer of these charge carriers across the p-n-boundary in a semiconductor detector. Especially at low light intensities quanta are observed at infrequent random intervals. It is obvious that current thus generated is "granular", i.e. significantly affected by noise. Along with thermal noise shot noise is classified as a form of white noise. The fluctuation of the "steady" current I_{DC}, i.e. the rms shot noise current I_n, may be calculated using the formula

$$I_n = \sqrt{2 I_{DC} e \triangle f} \,, \qquad (10.10)$$

where e is the charge of an electron. This equation assumes that the charges generating the current do not interact with each other as would be the case with e.g. a photodiode. An opposite example involves metal conductors in which a long-range correlation between the charge carriers may often be shown to occur. According to (10.10) current fluctuation I_n may be reduced by narrowing the frequency band or lowering the current. However, it should be noted that in the case of the latter operation the relative proportion of shot noise does not fall – quite on the contrary, it actually grows. For example, in a DC current of one ampere the level of noise is about 0.000006% while one pico-ampere contains some 5.6% noise over a bandwidth of 10 kHz.

10.3.3 Flicker Noise

The origin or generation mechanism of the ever-present flicker noise is not precisely known. It is, however, known that the magnitude of this type of noise increases as the frequency of the observed signal decreases. As a consequence of this characteristic it is also known as *1/f-noise*. This noise has been shown to occur in numerous natural phenomena. Indeed, neither the great currents of the oceans nor the delicate flow of sand in an hour glass are free from flicker noise.

Generally speaking, flicker noise becomes significant at frequencies of under 100 Hz. In addition, its effects may be reduced through the choice of a suitable type of resistor. Wire-wound and metallic film resistors are the most widely favoured in this respect.

10.3.4 Environment Noise

Environment noise includes all disturbances which originate outside the device itself. One significant source is the electromagnetic radiation which surrounds us everywhere. This is easily transferred to unshielded equipment, especially to electric wires (which behave just as antennae). Nowadays the list of sources of electromagnetic radiation is long. Manmade contributions include alternating current power lines, electric motors and radio and TV broadcasting antennae. Of natural sources we might mention the sun. Other phenomena beside environmental electromagnetic radiation also cause noise. Good examples are the daily and annual fluctuations in temperature on the earth's surface. In addition, mechanical vibrations such as sound waves may be connected to microphonic parts such as detectors and wires.

The frequency distribution of environmental noise resembles 1/f-noise in that its intensity per cycle increases as the frequency decreases. The frequency distribution also includes narrow and intense noise peaks originating from power lines and radio/TV broadcasting masts. From the entire spectrum of environmental noise we may find few regions where noise is at an acceptably low level. A relatively good region is that between 1 and 500 kHz.

This represents a “quiet band” in between those dominated by electric power lines and AM radio frequencies. Another disturbance-free zone is located at about 10 Hz just at the bottom end of the power cable range.

10.3.5 Stray Light

It has been said that of all the parameters in spectrophotometry *stray light* has caused perhaps the most confusion. The term stray light is used to describe light travelling along the optical path of the device but possessing a frequency or wavelength other than that which the user has intended. The mechanism of creation of stray light often relates to the use of a monochromator. In such cases it is referred to as *monochromator stray light.* In practice, stray light is generated via scattering and diffraction by the imperfect surfaces inside the monochromator. Furthermore, all gratings produce some degree of stray light although modern holographic gratings are an improvement on the traditional mechanical type in this respect.

The other parts and components of a spectrophotometer such as the detector tend to modify any stray light present. Stray light thus altered is known as *instrumental stray light.* This may be defined as the signal produced by the detector in response to radiation of wavelength which does not match the monochromator passband with respect to the total signal for any given wavelength setting [215].

Instrumental stray-light depends not only on monochromator stray-light but also the absorption of the sample and the response of the detector. Although the intensity of stray light is small at a set wavelength its effect on the final result may be significant because it adds up over the entire operational wavelength range of the detector. Especially noteworthy are cases where only a small amount of light of the desired wavelength arrives at the detector. This may result from e.g. strong absorption in the sample or a low level of emission from the light source. On the other hand some of the stray light arrives at the detector unattenuated as a result of its broad spectral distribution. This problem often appears in the UV range. Additionally, the sensitivity of the detector in the desired wavelength range may be low in comparison with its response to stray light. It should be noted that the sample too may act as a generator of stray light (e.g. fluorescence).

The relative proportion of stray light in a typical small monochromator is of the order of 0.005%. Stray light may effectively be eliminated by using two monochromators in series. The presence of a second monochromator reduces the amount of stray light by a factor of about 1000. Other procedures for reducing the stray light level and its effect on photometric accuracy have been studied e.g. by Slavin [216].

10.3.6 Chemical Noise

The chemical nature of a sample, for example, the position of chemical equilibrium and moisture content, depends on its chemical composition and structure. Chemical noise arises when some uncontrollable quantity changes the chemical nature of the sample. Generally, the uncontrollable variable arises in the environment. For example, temperature, pressure and air humidity may all change during the course of measurement while the operator remains either indifferent or unaware. These events may cause changes in the sample. In addition to these above-mentioned variables external light causes chemical reaction in light-sensitive materials. Similarly, shocks and other vibrations may alter the structure of the sample e.g. increasing the density of a powder sample.

10.4 Methods for Improving the Signal-to-Noise Ratio

In demanding and precise analyses and measurements a high signal-to-noise ratio is required. A good S/N ratio is not a foregone conclusion. Often a number of operations and solutions are required in order to achieve it. Also in cases where the signal itself is weak and of the same magnitude as the noise there is good cause to apply procedures which improve the S/N ratio. Over the years numerous methods have been devised for eliminating noise. These may be divided into two groups, hardware methods and software methods. In hardware methods various additional devices and components such as filters, modulators and electric shields are used to prevent noise attaching itself to the signal. Software methods operate according to a different principle, namely, the cleaning up of the already contaminated signal using mathematical means.

10.4.1 Hardware Methods

As we have already seen environmental noise is easily transferred to unshielded electronic components and wires in devices. For this reason disturbance-sensitive electronic components should be shielded by surrounding them with a conducting material which is attached to earth. Particularly sensitive components include high-resistance transducers in which even the smallest of current changes causes large voltage fluctuations in the output. Powerful transducers are generally located at the head of the amplification chain (e.g. a photodiode connected to a current/voltage converter). In such cases noise arising appears greatly augmented at the end of the amplification. In addition to shielding, the physical size of electronic components should be reduced as far as possible while effective earthing should also be ensured. Satisfactory shielding is often only achieved after numerous trials.

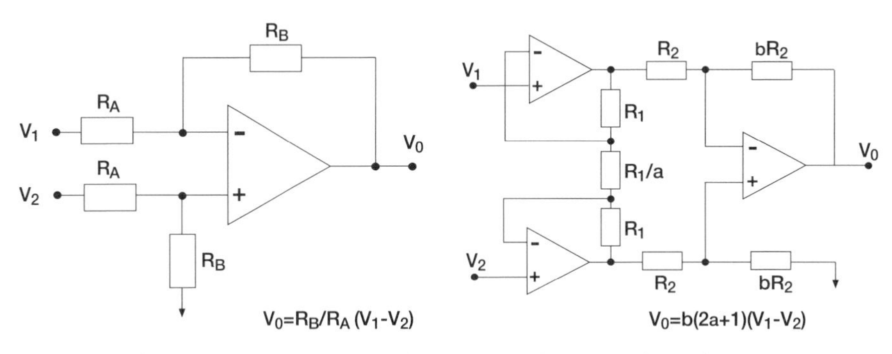

Fig. 10.9. An instrumentation amplifier may significantly reduce the noise common to both inputs

Managing the start of the amplification chain is thus of great importance in reducing noise. Amplification methods have, indeed, been developed which account for the presence of noise. These included, amongst others, *difference amplifiers* or *instrumentation amplifiers* as shown in Fig. 10.9. These are able to remove the common-mode-noise occurring in both inputs of an amplifier even by factors of 10^6. Analog electronics and its applications may be found e.g. in [217].

As has already become apparent the amount of noise and the frequency band are related to each other. In other words, as expressed in [214], "how much noise you see depends on how fast you look". As a result one method of combating noise is simply to restrict the frequency band. The transformation of the frequency distribution of an analogue signal may be performed using quite a basic frequency filter. If our signal frequency is low and possessing a narrow bandwidth we can exploit the characteristics of a *low-pass filter* to damp the voltage amplitude at higher frequencies. At the same time thermal and shot noise will also be damped. Correspondingly, if the instrument employs a high frequency we may reduce drift and flicker noise by using a *high-pass filter*. When the characteristics of low-pass and high-pass filters are combined we have what is known as a *band-pass filter*. This kind of filter is designed so that its minimum damping matches the frequency of the signal.

Flicker noise is most prevalent at low frequencies and in DC signals. It may be reduced by raising the frequency of the signal coming from the transducer. This process is known as *modulation*. Following modulation the signal may be amplified as required and fed through a high-pass filter to further remove noise. The signal is then returned to its original frequency by the converse process of demodulation and feeding through a low-pass filter. Modulation should be performed as close as possible to the signal source.

Modulation may be performed directly into the quantity to be measured, e.g. to light intensity or to the electric signal coming from the transducer. Modulation of a beam of light may be performed using e.g. an electric or mechanical chopper. For example, a light beam aimed at the sample may first be directed through a rotating and perforated disc. After passing through the chopper the intensity of light varies over time more or less as a square wave (its exact form depends on the dimensions and forms of the light beam and the aperture of the chopper). Electric signals such as voltage signals may be modulated by earthing the signal at suitable points along the amplification chain. Modulation at the desired frequency is performed using a solid-state switch.

The *Boxcar integrator* is a device used for removing impurities from the signal waveform and thereby for improving the S/N ratio. In practice this is achieved using a rapid acting electronic switch which collects analogue signal samples over the desired timeframe. Boxcar integrators are generally used with relatively fast signals, i.e. of time scale 10^{-6}–10^{-12} s. It is especially suitable for the observation of rapid physical and chemical phenomena which are excited by pulse lasers.

One of the most effective ways of modulating and removing noise is provided by the *lock-in amplifier*. Correctly applied this device is able to separate the desired signal from noise in cases where the noise intensity significantly exceeds that of the signal. The functioning of lock-in amplifiers requires a reference signal of the same phase and frequency as the signal to be measured.

10.4.2 Software Methods

Nowadays in the age of the computer signals may be cleaned by programs designed for the purpose. Also many of the hardware methods described above may be performed using appropriate software. We noted earlier that averaging is an effective and commonly used method for removing random noise from a signal. For example, a number of reflectance or transmittance spectra may be measured in succession and stored separately in the computer memory. If the measurement needs to be performed rapidly measurement data may be collected and summed together in a memory vector using capacitors in series (hardware averaging). This process is known as co-addition. Following measurement the values stored in the memory are divided by the number of measured spectra. The signal-to-noise ratio increases in direct proportion to the square root of the number of measurements. Thus, if we perform n measurements of a DC signal S then

$$S/N = \frac{\overline{S}\sqrt{n}}{\sqrt{\sum_{i=1}^{n}(\overline{S}-S_i)^2}} \,. \tag{10.11}$$

In the above formula $\overline{S}$ is the average of individual measurements S_i. An example of the technique of averaging is presented in Fig. 10.10. The curves in the diagram represent light reflected from a liquid (a voltage ratio) as a function of angle of incidence. In this test the measurement wavelength was deliberately set in the UV range at the limit of the useful wavelength range

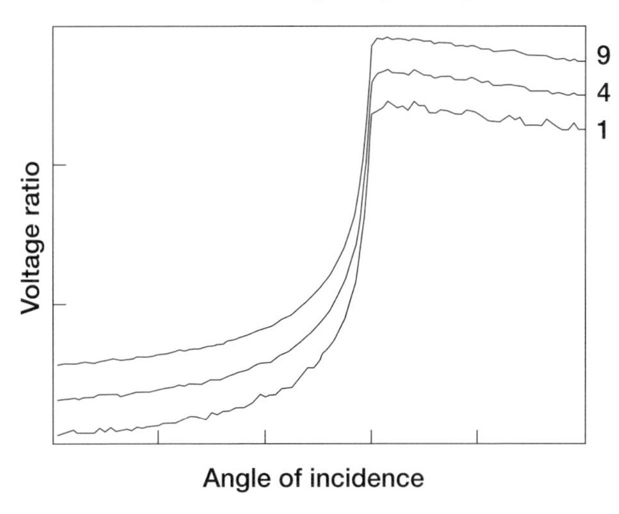

Fig. 10.10. Reflectance from glass/liquid interface as a function of angle of incidence (for clarity curves are positioned to different levels). The curves are averages of 1, 4 and 9 measurements. Noise is clearly reduced by increasing the number of measurements

so that the S/N ratio would be poor. Reading from bottom to top, the curves are averages of one, four and nine measurements respectively. The diagram shows clearly that averaging tends to reduce noise.

In practice, increasing the number of measurements makes the overall measurement process more time-consuming. An alternative way of reducing noise is to perform a single measurement over a greater period of time. Figure 10.11 shows the effect of increasing measurement time on S/N ratio.

Boxcar averaging may also be used with stored data (cf. boxcar integrator). Averaging is based on the principle that the signal changes slowly over time and that the signal at any randomly chosen moment in time is best represented by the average signal over a period preceding and following the moment in question. Problems may thus arise if the signal is complex and changes quickly. However, this method may be used if we require only the average amplitude of a periodic wave such as a square-wave.

Using the *Fourier transform* a signal which has been measured in the time domain may be transformed into a signal in the frequency domain. This transformation may also be performed in the opposite direction and into other variable besides time. In general, if the function f is dependent on variable β then its Fourier transformation $g(\alpha)$ may be written in the form [218]

$$g(\alpha) = \frac{1}{\sqrt{2\pi}} \int_{-\infty}^{\infty} f(\beta) e^{-\mathrm{i}\alpha\beta} d\beta \,. \tag{10.12}$$

Equation (10.12) is applicable to functions but not to the handling of finite number of measurements. Numerical calculations may be performed us-

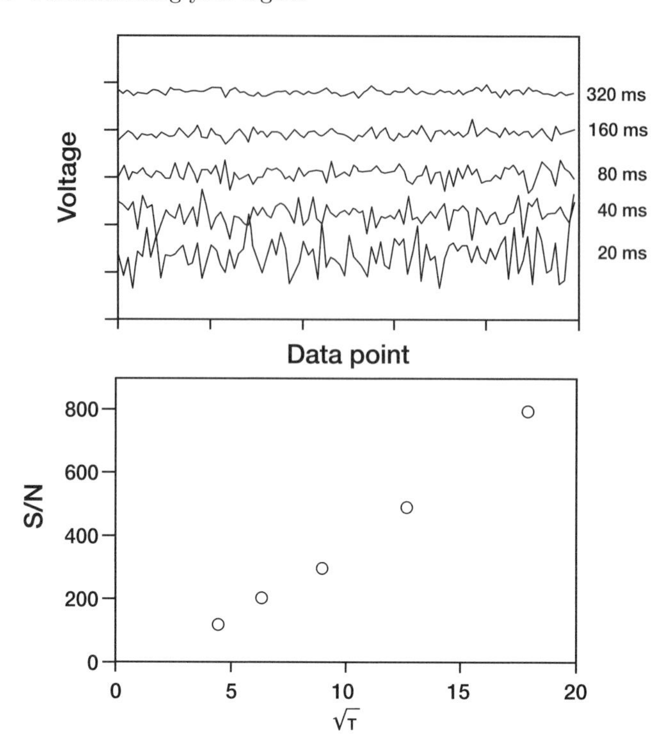

Fig. 10.11. A voltage signal consist of 100 data points. Each point in the sample is an average voltage of 20, 40 ... 320 ms measurements. The S/N ratios of voltage samples were calculated using (10.6). The result shows an almost linear relation between the S/N ratio and the square root of measurement time

ing quick and effective discrete algorithms developed for the purpose. Figure 10.12 demonstrates how the Fourier transform may be used to eliminate noise. In the example presented it is assumed that the signal itself changes more slowly over time than the noise. It follows from this that the useful information is located in the frequency domain in the left hand side of Fig. 10.12b while noise is mainly located at the right hand edge of the same diagram. We may now set the noise boundary at a given frequency f_{lim} and zero the transformed signal $g(f)$ from this boundary onward. Using the inverse Fourier transform we may return the signal from which the high frequency noise has now been eliminated to the time domain.

One very common smoothing method is to identify from the data obtained a part of the spectrum, the average of which represents a point on the smoothed spectrum. If the part of the spectrum in question has e.g. five points then we speak of *5-point un-weighted data smoothing.* Averaging may also be performed by weighting the points in the part spectrum. Smoothing is

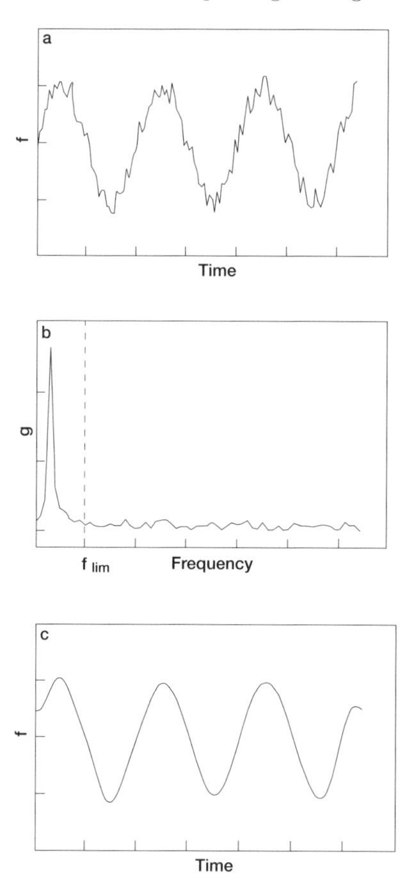

Fig. 10.12. (**a**) High frequency noise is added to a sinusoidal signal. (**b**) The contaminated signal is converted into frequency domain using the Fourier transform method. Noise may be eliminated by zeroing the function g at frequencies higher than the f_{lim}. (**c**) A clean signal is obtained after the inverse Fourier transform

more effective if we increase the number of data points in the part spectrum. Smoothing may, of course, be performed several times.

Data or spectral smoothing may also involve the use of a polynome. Here a part of a spectrum is fitted to the polynome using the least-squares method. The central point of the adjusted polynome curve now represents a point in the smoothed spectrum. This method is known as *least-squares polynomial data smoothing* and it functions in much the same way as a low-pass filter. However, in cases involving large quantities of data it may require considerable time and calculation power.

Savitzky and Golay [219] have developed a smoothing method in which certain integers known as *convolution integers* may be used as weighting coefficients in data smoothing. In fact, it may be shown that the use of such

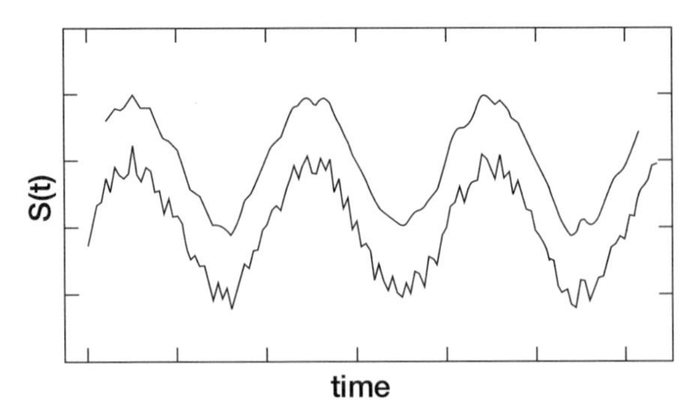

Fig. 10.13. Noise from sinusoidal signal (lower curve) is removed by Savitzky–Golay method. Here 5-point quadratic smoothing function used convolution integers: -3, 12, 17, 12, -3 [219]

weighting coefficients corresponds precisely to least-squares polynomial data smoothing. An example of the *Savitzky and Golay method* in practice is given in Fig. 10.13 in which sinusoidal signal is contaminated with noise. In this particular case the points of the curve are processed twice using a 5-point quadratic smoothing function.

It should be noted that all data smoothing processes restrict the frequency band of the signal and therefore to some extent also distort the true signal. The effectiveness of smoothing methods in improving S/N ratios is also rather limited. However, the use of such methods may be justified in that they produce a spectrum the features and details of which may be examined visually.

Correlation methods are based on complex mathematical processing which is performed either by computers or advanced analog instrumentation. The method is used e.g. when searching for a signal hidden by noise, for distinguishing overlapping absorption peaks in spectroscopy and chromatography, and in the smoothing of noisy data.

11 Before Measurement

Knowledge about physical and chemical phenomena is based upon observation and measurement. Phenomena are described using various quantities such as length, time or temperature. The base units of these quantities are given by the SI (Système International) system. Each quantity sets its own requirement of measurability. In general, the quality of a measurement result depends on the use of an unbroken chain of calibration and the adoption of internationally recognised measurement norms.

In this chapter we shall examine the most common routines, standards and tests by which instrument parameters and output values may be reliably combined with measurement quantities. This process may be termed calibration and involves e.g. the determining of a calibration curve as described in the preceding chapter. With the help of various tests we may also check that the spectral device functions as it is intended to. Only when we are certain of this may we be confident about the results it produces. Although the majority of literature has tended to concentrate on instruments measuring transmittance, most of the methods described below may also be applied to the calibration of reflectance measurement. Our survey emphasises the most practical methods while further details are provided in references [220, 221].

The history of photoelectric spectrophotometry began in the 1940s and the first devices developed were capable of reasonably accurate quantitative measurements. However, a serious problem arose as spectrophotometry became more widespread. Results from different laboratories often deviated significantly from each other. Founded in England in 1948, the Photometric Spectrometry Group set out to confront and solve these and various other problems. Later on as the aims of the group broadened its name was changed to the Ultraviolet (UV) Spectrometry Group. The group has published numerous important studies on absorbance and wavelength standards, stray light and other related subjects.

Over the years numerous calibration methods have been developed for use in UV/visible spectroscopy. Good methods are those which are easily applicable and available, suitable for the instrument in question, unaffected by environmental conditions and safe to use. In addition, the performance of the test should resemble the actual measurement of the sample.

11.1 Absorbance Standards

Absorbance may be checked using test samples or series in either the solid or liquid phase. Although the optical neutrality and temperature coefficient of liquid calibration samples do not reach the standards of solid test samples liquids have two important advantages. Their use closely resembles actual measurement (indeed, this book is concerned with a measurement of liquids) and they may usually be prepared by the user. However, their preparation demands care and accuracy. In addition, following preparations solutions require a specific storage time before use.

11.1.1 Calibration Liquids

The use of potassium dichromate as a calibration agent is referred to in numerous studies. Although, this compound is very useful the solvent must be carefully chosen. Research has shown that in the 200–400 nm range best results are achieved using 0.01 N sulphuric acid. The pH value of the solution is of critical importance (pH=3 being a good choice). Potassium nitrate is also suitable for the UV range. When solvated by water it produces a broad absorption peak centered at 302 nm. However, there is no precise linear relationship between the concentration of the solute and the absorbance measured. In other words, potassium nitrate does not obey Beer's law. Pyrene forms a very stable solution when solvated by iso-octane. Its spectrum contains numerous sharp absorption peaks which may be used to expose dynamic errors. One disadvantage is that the solution has a tendency to fluoresce. Compounds such as potassium hydrogen phthalate, picric acid and nicotinic acid may be used given certain restrictions.

Calibration standards used for the visible range include Thomson's solution, cobalt (II) ions, cobalt ammonium sulphate, nickel (II) ions, copper (II) ions and iron-dipyridyl complex. The most useful of these is cobalt ammonium sulphate which may easily be made into solution.

The NBS "composite" solution may be used across the entire UV/visible range. After careful ageing the solution remains stable and does not fluoresce. Its spectrum includes several broad absorption maxima and minima within the 250–600 nm range. The use of organic dyes such as Neolan Black, Cibalan Black, Alizarin Light Grey, Luxor Fast Black L and Nigrosine in absorbance calibration has also been closely studied. The problem with these commercial dyes is that they contain significant levels of impurities. The study of linear spectrophotometric operations has revealed the suitability of the colouring agent known as French's Green Food Colouring.

Liquid calibration standards are provided by recognised institutes and laboratories of standards as well as number of instrument manufactures [222].

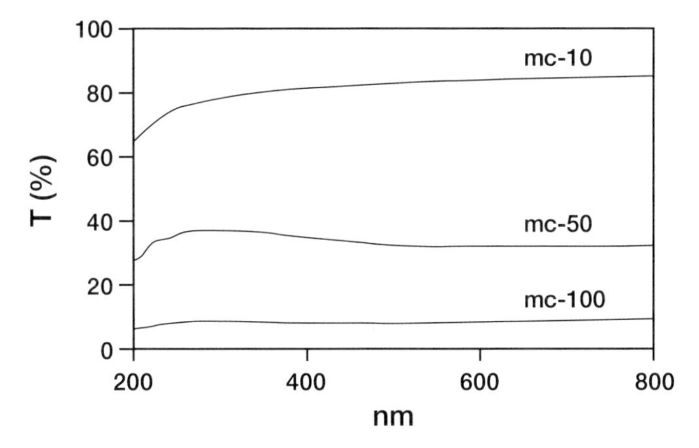

Fig. 11.1. The transmittance curves $T = T(\lambda)$ of three calibration solids (labelled as mc-10, mc-50 and mc-100). These "chromium-on-quartz" filters offer a useful tool for inspecting absorbance

11.1.2 Solid Standards

Solid phase materials may also be employed when checking absorbance and the linear operation of a device. Here calibration is effected with the help of polarizers or glass or metal filters or by controlling the light beam to the detector using openings and screens. Sets of neutral Schott glass filters are widely used in absorbance calibration. These are supplied by several instrument manufacturers as well as by recognised standard laboratories [223].

Generally the front and rear surfaces of the glass filters have to be parallel and should be fitted with special holders to prevent mechanical stress on the glass. Unfortunately glass filters cannot be used in the UV range as they tend to absorb such wavelengths. All standards undergo small changes during time. Therefore, glass filters should be recalibrated at yearly intervals.

UV range calibration may be performed using fused silica or quartz onto which metal vapour has been condensed into thin layer. The thickness of the metal layer determines the degrees to which light is transmitted and reflected. As this method is not based on absorption the reflected light (and afterwards scattered light) may reduce the accuracy of the calibration. Examples of transmission curves for chromium-on-quartz filters (SRM 2031) are given in Fig. 11.1.

The light power reaching the sensor may be controlled using polarizers. The light transmitted through two linear polarizers depends on the mutual orientation of their transmittance axes. In the system presented in Fig. 11.2 three polarizers are used – the third polarizer serves to eliminate polarisation caused by the optical and other components of the spectrometer itself. The transmission axis of the outer polarizers P_1 and P_3 are aligned in parallel. The central polarizer P_2 regulates light intensity. The polarizers should be of high

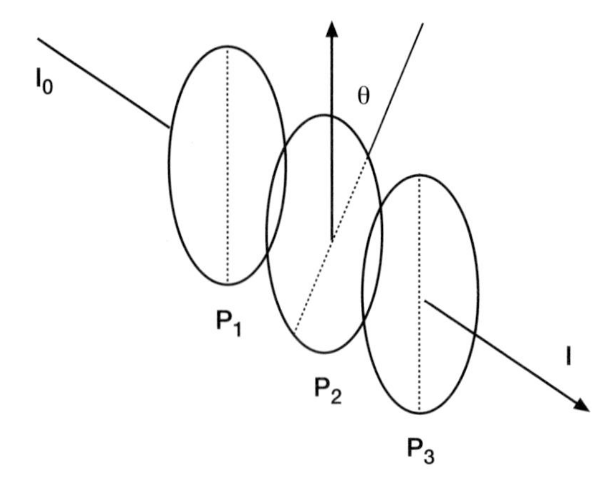

Fig. 11.2. A system of three polarizers for linearity checking. When the transmission axes of P_1 and P_3 are parallel the polarizer P_2 may be used to control the transmitted intensity

quality and should have good extinction ratios. Similarly, the rotation of the central polarizer should be carefully controlled. Given these requirements, the linearity of the method is better than 0.1% over an absorbance range of 0–4 AU (absorbance units) [224]. Other types of polarizers besides linear polarizers may also be used. However, in general the use of polarizers is less straightforward than that of glass filters.

The linear functioning of the device may also be tested using the summing method. This rather old method involves the use of two incoherent light beams. The device functions linearly when the sum of the intensities of the two light beams measured separately is the same as the intensity of the two beams measured simultaneously.

11.2 Monochromator Slitwidth

The width of the output slit of a monochromator is of great importance in UV/visible spectroscopy as it is used to control the spectral resolution or bandwidth required in the given measurement. Also the output power is affected by the slit. However, the width of the slit and the output power are not necessarily directly related – it also depends on image of the light source on the slit and on the source spectrum. The intensity distribution of the emerging band with respect to wavelength generally resembles a triangle as shown in Fig. 11.3. The peak of the triangle, i.e. the intensity maximum, corresponds to the nominal wavelength λ_0. The width measured between the two points of half maximum height of the triangle (FWHM) is known as the *effective spectral slitwidth*, ESW. Furthermore, the width of the spectrum which images onto the exit slit of a monochromator is defined as the *spectral slitwidth*,

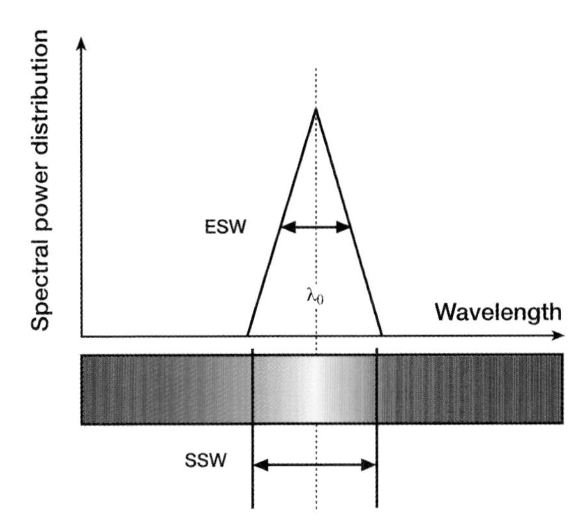

Fig. 11.3. The spectral intensity distribution of light exiting a monochromator often resembles a triangle centred at the set wavelength. The effective spectral slitwidth (ESW) corresponds to the FWHM of the triangle. The output slit of a monochromator determines the wavelength range (SSW) which is allowed to pass through the monochromator

SSW. We may also define the *natural bandwidth* (NBW) of the absorption peak as the FWHM value measured by a device of infinite resolution [225].

It is obvious that when radiation with wavelength differing from the nominal wavelength arrives at the sensor the final result will be distorted. Thus, for example, the measured absorption peak may be lower than the true value. In this case the ESW should be reduced to as close to zero as possible in order to minimise the bandwidth. In practice, however, such a procedure cannot be performed because as the slitwidth narrows so less and less light reaches the sample and the detector. This in turn reduces the S/N ratio. A compromise value for the monochromator slitwidth must therefore be found, i.e. a value which produces the required S/N ratio without significantly distorting the spectrum.

In critical measurements the user is well advised to try out the effects of different slitwidths on the maximum absorption peak values of the sample in question. It is often the case that on narrowing the slitwidth a point is reached where the absorbance no longer changes. This slitwidth may then be used for measurement. If, however, we have some knowledge or estimate of the natural bandwidth of a particular absorption peak we may use as a rough guideline the rule that the SSW should not exceed a tenth of the value of the NBW [225].

Benzene vapour (absorption peaks) or the deuterium lamp (emission peak at 656.1 nm) are simple yet useful means for checking the resolution of a spectral instrument.

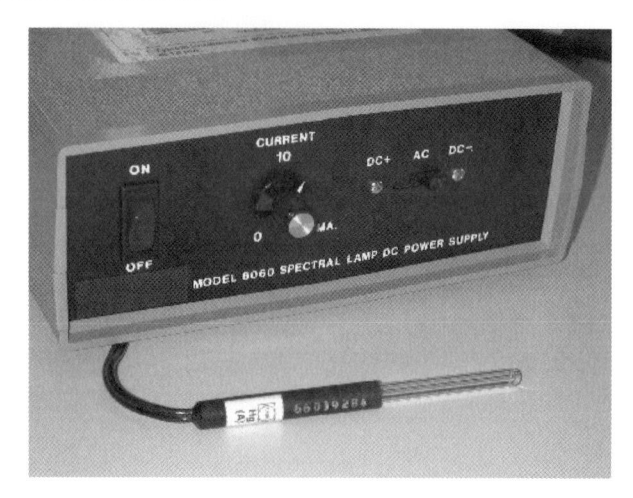

Fig. 11.4. A pen calibration lamp with power supply for wavelength calibration. This Hg(Ar) lamp produces a number of useful lines in the UV/Vis range

11.3 Wavelength Calibration

The wavelength accuracy of spectrophotometers intended for general laboratory use is typically better than 1 nm. Furthermore, the wavelength precision is generally one order of magnitude better than this. Although the wavelength stability of devices has improved over the years it should be checked periodically. Usually in wavelength calibration the instrument or some accessory device attached to it is used to produce an emission or absorption band which is sufficiently narrow to permit accurate calibration. Wavelength is adjusted (using software) if the wavelength difference between the measured and true values is greater than a particular set limit.

One accurate wavelength calibration method involves the use of low pressure gas-discharge lamps. These are used in the testing of e.g. monochromators, spectrographs and spectral radiometers. Such lamps produce narrow emission bands, the wavelengths of which depend on the gas, metal vapour and internal materials used in the lamp. Typical materials include Hg(Ar), Hg(Xe), Xe, Ar, Ne, Kr.

A requirement for the production of a narrow band is the low pressure of the gas inside the lamp. In normal laboratory conditions lamps may be relied upon to accurately produce their intended wavelengths. Only in cases of large changes in temperature or the presence of strong electric or magnetic fields is there the danger of the wavelength shifting. The wavelength accuracy of calibration lamps and the stability of their spectral irradiance have been reported by Sansonetti et al., amongst others [226].

Figure 11.4 presents a pen calibration lamp and power supply. The useable spectral lines produced by this Hg(Ar) lamp are given in Table 11.1.

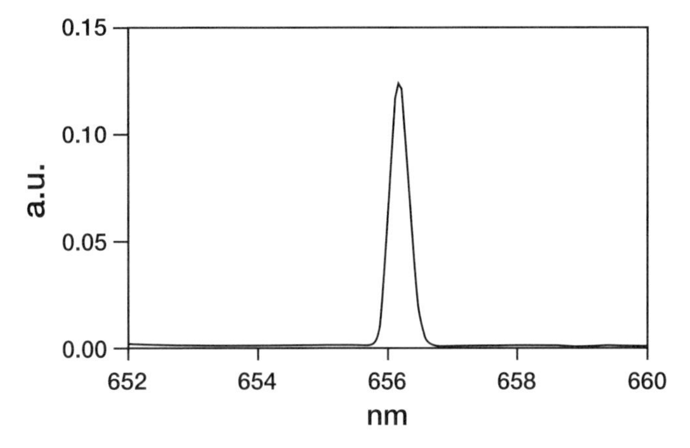

Fig. 11.5. A deuterium lamp emits a narrow line at 656.1 nm. Since most spectrometers employ deuterium lamps wavelength calibration is easy to perform

In practice, the calibration lamp should be installed directly in front of the monochromator input slit.

Table 11.1. Useable wavelengths (nm) of Hg(Ar) lamp [227]

184.9	187.1	194.2	253.65	265.4	284.8	302.2	312.57	313.15	313.18
320.8	326.4	345.2	365.02	404.66	435.84	546.07	576.96	579.07	615.0

The outputs of many broadband light sources also contain useful emission peaks. For example, Hg(Xe) lamps produce numerous points of reference in the visible range, the true wavelengths of which may be compared with the values produced by the spectral instrument. The output of a deuterium lamp lies mostly in the UV range and diminishes on approaching the visible spectral range. Nevertheless, the lamp provides a useful narrow band at 656.1 nm (Fig. 11.5). Many spectrophotometers actually employ deuterium lamps to produce UV radiation and therefore the wavelength calibration of such a device using a deuterium lamp is particularly convenient.

A suitable filter may also be placed across the light path in the sample chamber. This produces one or more absorption bands which may be used in wavelength calibration. Unfortunately the repeatability achieved in filter manufacture has fallen behind in comparison with the general level of development in the field of spectrophotometry. However, such filters are well suited for daily routine examinations. Calibration over the entire UV/visible range using a single filter may be carried out using e.g. a holmium or didymium filter [228]. The transmittance of a holmium filter is presented in Fig. 11.6. Also solutions containing holmium (III) ions may be employed. In exceptional

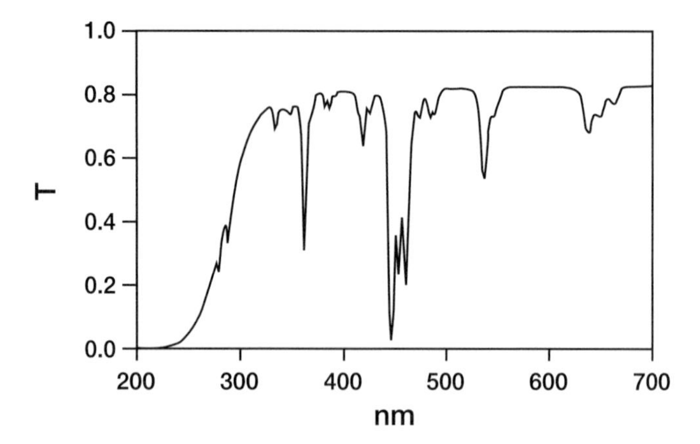

Fig. 11.6. The transmittance curve of a holmium filter

circumstances a narrow-band interference filter may be also used although these produce only a single calibration point per filter.

11.4 Angle Calibration

The functioning of the device may involve rotational movements the orientation of which is of vital importance for correct operation. The calibration of rotators in commercial instruments takes place automatically when the power is switched on. However, this may not be the case if the operator uses self-made spectroscopic devices. Some rotators and associated control units have built-in zero-angles or reference points. If this feature is not available then angle calibration must be performed by some other means. In a reflectometer built by the authors (see Chap. 12) angle calibration was achieved using a laser, a mirror attached to a rotator and a slit-detector (Fig. 11.7). When the rotator begins to rotate a voltage curve is traced over a narrow arc as shown in Fig. 11.8. The maximum of the curve may be taken as the reference point of the rotator. In this particular case the reference may be determined to an accuracy of 0.01 degrees of arc.

11.5 Baseline Checking

Spectrophotometric measurements are often relative, i.e. we are only interested in measuring the light attenuation caused by the sample with respect to some reference level. The reference level is determined by measuring some suitable reference sample. This may be water or some other solvent or even air. When the same material is used as both reference and sample the result produced by the spectrophotometer is known as the *baseline*. Baselines

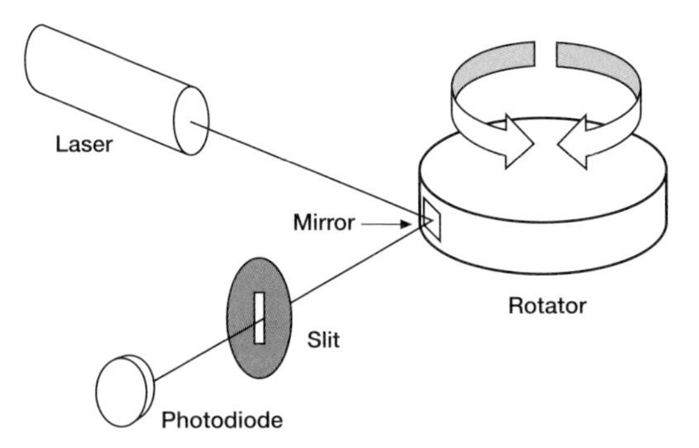

Fig. 11.7. An angle calibration system employed with the reflectometer developed by the authors. The system includes a HeNe laser, a slit, a photodiode and a small mirror attached to a rotator

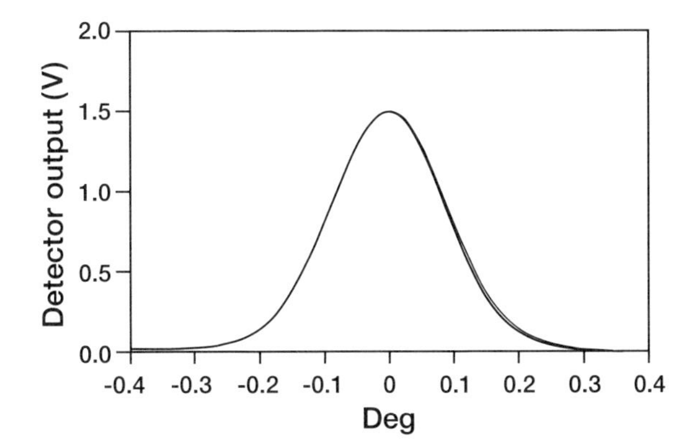

Fig. 11.8. When the rotator rotates the photodiode plots a Gaussian shaped voltage curve, the maximum of which may be considered as the reference angle

may be obtained for either transmittance or reflectance. A perfect instrument would produce a 100% or, alternatively, zero absorbance reading over the entire measured spectrum. In practice, the measured baseline may also contain distortion. This is described by the term *baseline flatness*. It should be noted that the term flatness does not refer to the high frequency or rapid fluctuation of the baseline value which is caused by noise. Manufacturers specify the baseline flatness of their devices. Limits are typically of the order of ± 0.001 AU. Examples of baseline measurements are given in Figs. 11.9a and b.

The quality and stability of the instrument may also be estimated from the alteration in the baseline over time. This *baseline stability* is indicated in terms of the change of absorbance value over a given period of time at a particular wavelength. For example, the baseline stability of one commercial photometer is claimed to be better than 0.0004 AU per hour at a wavelength of 340 nm.

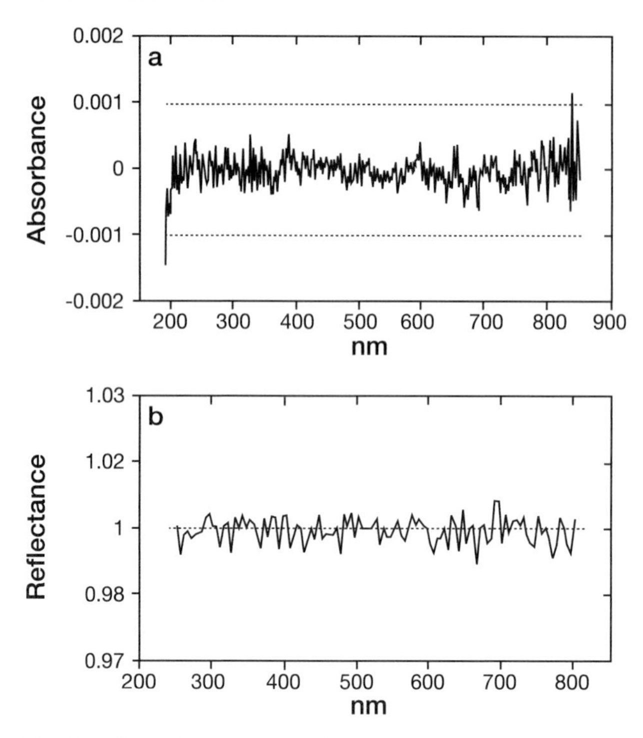

Fig. 11.9. The baseline of a commercial spectrophotometer depicted in terms of absorbance. The dotted lines (±0.001 AU) define the area in which the baseline should lie (upper figure). Respectively in the lower figure, a measured baseline of the reflectometer (presented in Chap. 12)

11.6 Recommended Methods

A number of routines exist which may be applied to spectrophotometric calibration. In choosing between them account should be taken of the accuracy required by the measurement. Other factors to consider include the availability, manufacturability and stability of the calibration device. Listed below in Table 11.2 are procedures recommended by the UV Spectrometry Group [220]. The concept of stray light is discussed in Sect. 10.3.

Table 11.2. Recommended procedures for calibration

Parameter	Calibration Tool	Attribute
Monochromator resolution	Bentzene vapor	Peak at 259.6 nm can be resolved if the spectral slitwidth, SSW is less than 0.5 nm.
	Deuterium lamp	FWHM of observed emission line at 656.1 nm corresponds to the SSW
Wavelength	Discharge lamp (Hg)	Several lines in UV/Vis range
	Solution of Holmium III ions	Several lines in UV/Vis range
Stray light (instrumental)	Aqueous cut-off filters: KCl, NaI, NaBr, $NaNO_2$	Useable range: 165 ... 385 nm
	Vycor glass	Useable range: 200 ... 210 nm, quick check
	Glass filters	For visible and NIR range
Absorbance (linearity)	Cells with difference pathlength.	Single concentration applied to cells with e.g. 5 to 40 mm pathlengths
	Solutions that obey Beer's law	e.g. green food dyes
Absorbance (accuracy)	Set of neutral glass, Nichrome-on-fused quartz filters	For high accuracy calibration
	Potassium dichromate in 0.01N sulphuric acid	For routine calibration

12 From Theory to Measurement

In this final chapter we shall present results obtained from the application of methods and theories described in Part I of this book. Actual measurements were performed using a reflectometer constructed for the measurement of liquids. Our main aim was to seek and develop effective measurement and calculation techniques for determining the optical properties of various types of liquids. By optical properties we mean those quantities which describe the progress of an electromagnetic wave through a homogenous liquid – in other words, the refractive index and the extinction coefficient. Our choice of samples also included diffusing liquids such as milk and paper coating suspension. Our interest lay chiefly with reflection measurements because, in terms of measurement geometry, these are well suited to e.g. problematic industrial process measurements.

12.1 Reflectometer

Let us first consider the main tool at our disposal, the reflectometer, since the measurements presented later on were performed using this device. The instrument in question is a spectral reflectometer specifically intended for the study of liquids. The functioning of the device is based on the observation and analysis of light reflected from the prism-sample boundary. There are two reflectance geometries from which to choose, i.e. bidirectional reflectance and directional-conical reflectance. The instrument was constructed on a small optical base plate mostly from readily available commercial components. The number of self-made components was kept to a minimum. An internal view of the measurement unit is presented in Fig. 12.1.

12.1.1 Optical Layout

The optical principle of the instrument in reflection mode is presented in Fig. 12.2. The light source chosen was a 75 W xenon lamp. This produces a relatively even output with the exception of a few strong emission peaks in the long wavelength end of the visible range and in the NIR. Light from the lamp is directed using a mirror into the input slit of a monochromator. The monochromator used was a grating monochromator (the operation of which

Fig. 12.1. A photograph of the measurement unit of the reflectometer

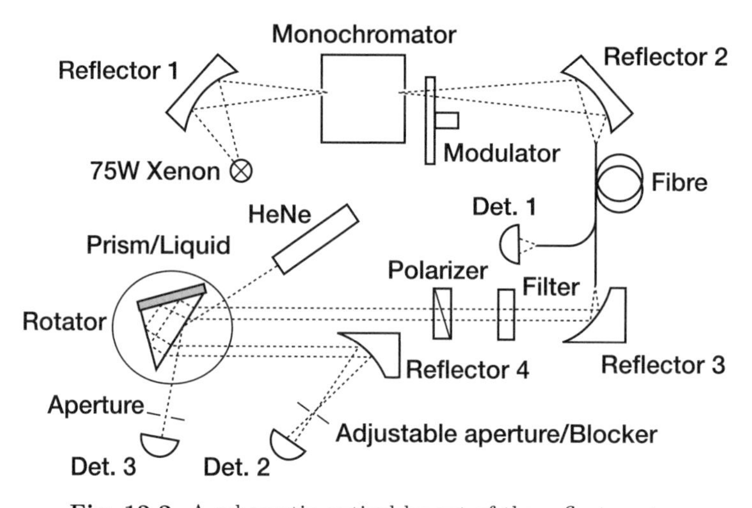

Fig. 12.2. A schematic optical layout of the reflectometer

is described in Sect. 9.2). In practice, the user may select any wavelength in the visible region ranging down to 250 nm in the UV. Wavelength setting or scanning may be performed either manually or automatically using software. The monochromator slit widths were chosen so as to correspond with a wavelength bandwidth of 4 nm.

Light exiting from the momochromator is directed at a fibre using a focusing mirror. The diameter of the fibre used was 100 μm and both the core and cladding were made of fused silica. The fibre is bifurcated, one of its branches being used for detecting changes in the intensity of the light source and the other for generating the measurement beam. This latter part of the fibre is placed at the focal point of parabolic reflector. This allows for the generation of an almost completely collimated beam. The beam is linearly polarized and directed through a prism to the liquid sample. The sample is fed into a sample chamber attached to the side of the prism using a pump. The sample may be either stationary or in continuous flow during measurement. The prism and attached sample chamber are mounted on a motorised rotator. This allows for the desired angle of incidence to be chosen or the scanning through a particular range of angles. The rotator in question may be rotated through an angle as small as $0.01°$. Calibration (zeroing) of the rotator is performed using a laser, a mirror attached to the rotator and a slit detector. Calibration is described in Sect. 11.4.

The other side of the prism was made into a mirror by coating its surface with a layer of silver. Light entering the prism experiences two reflections, i.e. from the sample and from the mirror. After these reflections it leaves the prism in a direction parallel but opposite to that of the incoming light. The beam is then directed by the focusing mirror to the aperture system located at the focal point of the mirror and then on to the main detector. The aperture system is made up of various sized openings machined onto a rotating disc. When a small aperture is selected the detector perceives mostly light rays which leave the prism collimated and parallel to the measurement beam. Since light scattered in the system is now effectively blocked from entering the detector we may speak of specular reflectance (bidirectional reflectance). On increasing the aperture size the reflectance geometry changes to that of directional-conical reflectance. Here the light entering the detector may possibly contain a scattering component. The aperture system may be used in yet a third manner. If a small circular-shaped barrier is positioned at the focal point then the path of the specular component is blocked. In this case mostly light which is scattered by the sample will be perceived. The measurement geometry here resembles directional-conical reflectance except that one particular direction of observation, that of specular reflection, is absent.

The detectors used in this instrument are commercially available silicon-based photodiodes. Voltage-current amplification is controlled using software, typical gain being in the order of 10^8 V/A. With the exception of the polarizer all the functions of the instrument are automatically controlled. Data processing and analysis is performed using separate mathematical software.

The instrument may also be used for traditional transmission measurements. Here a cuvette containing the sample is placed in the light path in front of the prism. The prism is made "invisible" by turning it to the area of total reflection (the sample in the chamber being water). The instrument now functions as a traditional single-beam spectrophotometer.

12.1.2 Measurement Modes and Parameters

The user may choose

- the wavelength and bandwidth of the measurement beam,

- the angle of incidence,
- the polarization angle of plane polarized light,
- the measurement of either specular reflection or scattered light,
- scanning of measurement parameters.

These make possible measurement of the following primary reflectance data

- $R = R(\theta)$, λ constant,
- $R = R(\lambda)$, θ constant,
- R_s and/or R_p, λ and θ constant.

Naturally, these may also be performed with respect to time when studying, for example, the contamination of the prism surface or other gradual processes. In addition, the instrument provides the opportunity for performing SPR (surface plasmon resonance) measurements simply by replacing the prism by one which has a thin metal film on the surface forming a boundary with the sample.

12.1.3 Primary Data of the Apparatus; Examples

The left hand side of Fig. 12.3 shows the reflectance curves of three dye solutions as functions of wavelength, i.e. $R = R(\lambda)$. In terms of their external appearances they were green, red and blue. The dye solutions were prepared by adding dye to water until the solution became saturated. The aim was, therefore, to produce highly absorbing samples. Wavelength scanning was performed across a range extending from 250 to 700 nm at 5 nm intervals. The angle of incidence was set at a constant value below the critical angle. It should be remembered that reflectance is highly dependent on the angle of incidence in cases where this is set just below the critical angle. For sake of comparison, the samples in question were diluted in water by a factor of one part in a hundred. The samples were then placed in a 10 mm cuvette and subjected to transmission measurements employing a commercial spectrophotometer. The results of this are also presented in Fig. 12.3. We note that deep transmission minima, i.e. absorption bands cause features which may be discerned in the reflectance curves (as, indeed, theory would predict).

The instrument may also be used to measure reflectance R in relation to angle of incidence θ as shown in Fig. 12.4. In this case the wavelength was kept constant and the angle of incidence varied through approximately 13° in the vicinity of the critical angle. Angle scanning was performed at 0.1° intervals. Figure 12.4 presents the reflectance curves obtained from measurement of the ethanol and lignin solutions. The former of these was measured at a wavelength of 589 nm and the latter at 280 nm. The ethanol solution is transparent ($k \approx 0$) in the visible part of the spectrum. Its transparency is particularly evident at the sharp folding point of the reflectance curve, i.e. at approximately 67°. Light reflected at angles greater than this critical angle experiences total reflection at the prism-ethanol boundary. The refractive

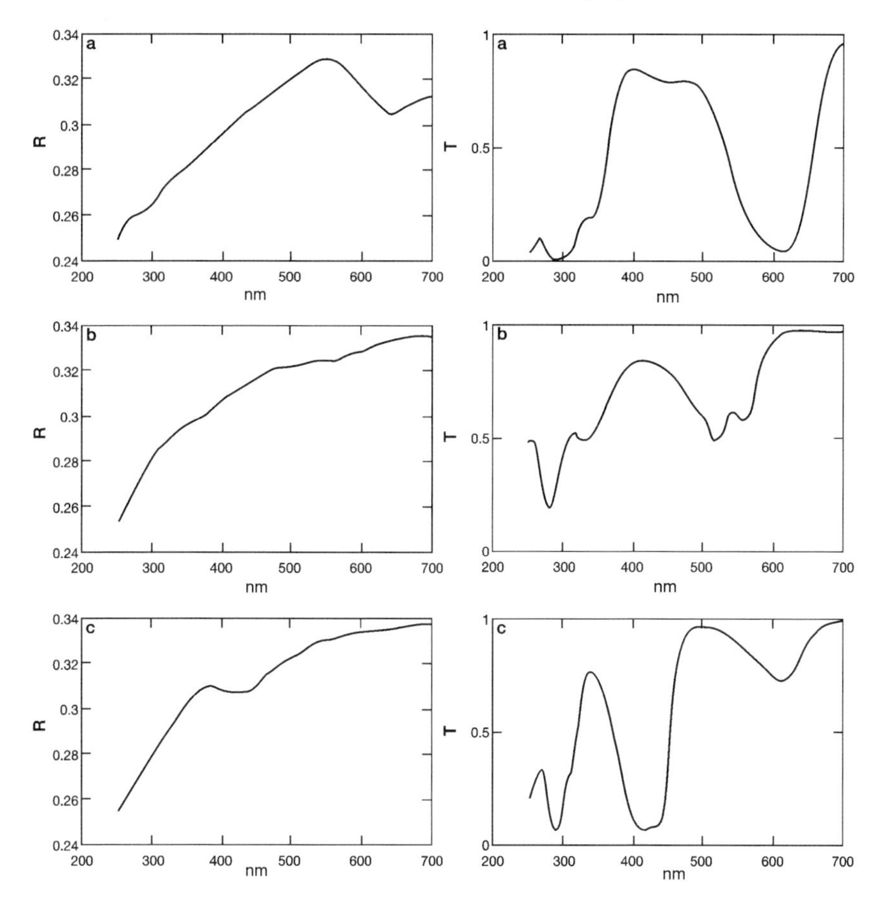

Fig. 12.3. The reflectance $R = R(\lambda)$ and transmittance $T = T(\lambda)$ curves of (**a**) blue, (**b**) red and (**c**) green dye solutions. For the transmittance measurement solutions were diluted (1 : 100) because of their high optical density

index of the ethanol solution at the wavelength used is 1.3428. By contrast, lignin is an organic complex which strongly absorbs UV radiation. This absorption reveals itself in a levelling of the R curve in the proximity of the critical angle. The complex refractive index of the lignin solution measured here was 1.3891–0.0054i at a wavelength of 280 nm.

12.2 Practical Examples of Determining Optical Constants

We have seen that, in principle, many theories and methods of calculation are available for determining the optical properties of liquids using reflectance

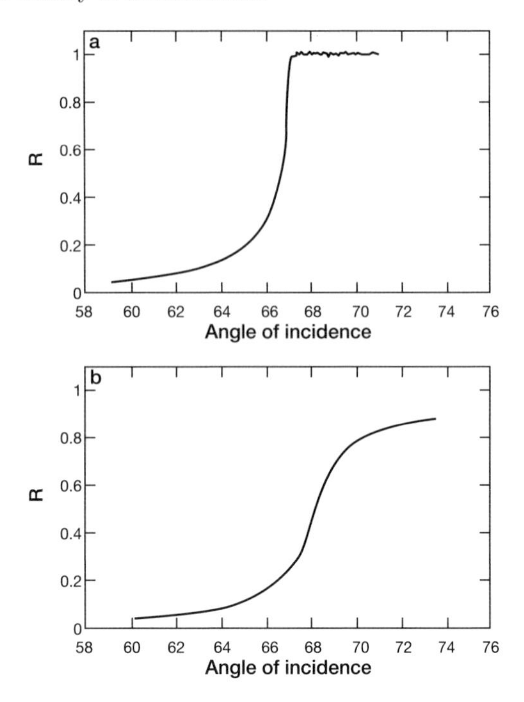

Fig. 12.4. The reflectance $R = R(\theta)$ curve of (**a**) ethanol and (**b**) lignin solution. The wavelengths used were 589 nm and 280 nm, respectively

measurements. These methods may be classified into two groups according to whether they produce a discrete result in the wavelength range or continuous spectral information. The first group includes the Fresnel fitting method, Azzam's polarization method and SPR measurement. Spectral information, on the other hand, is provided by MEM and ATR calculations. In fact, the Fresnel fitting method, for example, may be used to obtain refractive index and absorption spectra by measuring a large number of $R = R(\theta)$ curves at various wavelengths. In practice this takes up so much time as to cast doubt on the practicality and sense of the method. The choice of an appropriate method is also influenced by the degree of accuracy required.

12.2.1 Fitting the Reflectance Curve to Fresnel's Equations

Fresnel's formulas (4.4) and (4.5) combine the intensity of light reflected from the boundary of two homogenous materials with the refractive index of the materials. In the case of a single interface, i.e. two materials, the equations may be used in their usual form and experimental data may be fitted to them. When successful this yields the relative refractive index of the two materials n_1/n_2. The experimental system may be arranged in such a way that the optical properties of one of the materials (n_1) is known. This is, indeed, the case with our reflectometer since we have information of the dispersion characteristics of the prism used in it. Consequently, the refractive index n_2 of the second material, i.e. the material being studied, may be calculated. Fitting may be effected by, for example, minimising the sum S which is dependent on the refractive index n_2.

$$S(n_2) = \sum_{\theta} \left[R_m(\theta) - R_F(\theta, n_1, n_2)\right]^2 . \tag{12.1}$$

Here R_m and R_F are the reflectance curves obtained from experimental measurement and as calculated from Fresnel's equations, respectively. The minimising of the sum S is a non-linear task involving the least squares which may be performed by, for example, the Leveberg–Marquardt calculation. The examples in Fig. 4.7 demonstrate the usefulness of the fitting method. The circles in the diagram represent points on the experimental reflectance curves of ethanol and lignin solutions. The fitting calculation seeks a value for n_2 at which the experimental R curve and the curve calculated from Fresnel's equation (solid line) are as identical as possible. Reference tests using other refractometers produced comparable results. It should be noted that this method is also applicable when the liquid under examination absorbs light, as in the case of the lignin solution. Here fitting also gives the extinction coefficient k in addition to the refractive index.

The linearity and applicability of this method over a wide range of concentrations has been tested out using prepared series of solutions. Figure 12.5 presents the results from one particular test. Here the lignin solutions were measured using a reflectometer. Following this the complex refractive indices of the solutions were calculated using the fitting method under discussion. The concentrations of the solutions are indicated along the horizontal axes of the figure. It may be shown that in this case the relationships between the optical constants and concentration are highly linear throughout the entire concentration range.

The applicability of the fitting method to the analysis of homogenous solutions has been demonstrated in numerous experiments. The situation becomes more complicated when the solution contains large concentrations of solid impurities or particles. When this is the case the light beam refracted into the solution will be scattered by the particles and this scattered light will most probably find its way to the detector if there is no system of blocking to prevent it. Thus, a scattered component becomes attached to the specular component in the signal. Clearly, Fresnel's equation and the fitting method described above are not directly applicable in such a case. This was demonstrated by the following test. We set the aperture system of the reflectometer to its widest possible opening and measured the reflectance curve for a strongly scattering liquid, i.e. milk, using s-polarized light. The data obtained was processed in the normal manner using the fitting method. It is obvious that the fitting for a light-scattering liquid such as milk will not be satisfactory, as may be seen from Fig. 4.13a. If we describe the scattered

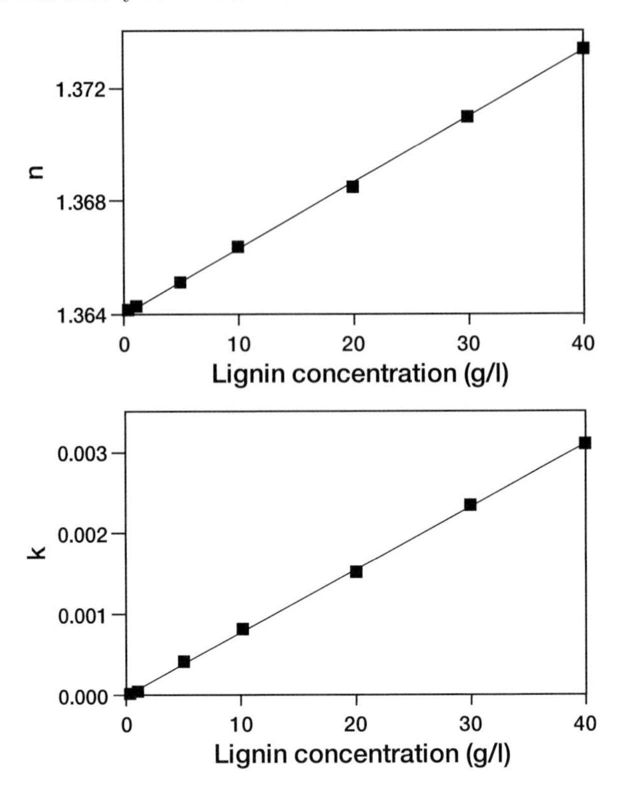

Fig. 12.5. The refractive indices n and the extinction coefficients k of various concentrations of lignin-water solutions determined using the fitting method

component of light arriving at the detector by a term which takes account of the angular dependence F of scattering, the unitless scattering parameter ϕ, and the proportion of light refracted into the sample $(1 - R_s)$, we obtain a scattering-corrected expression for reflectance as follows:

$$R_s^+ = R_s + F\phi(1 - R_s) \ . \tag{12.2}$$

Now, using this modified Fresnel equation, the fitting operation is more successful as Fig. 4.13b indicates. The calculation also provides us with information about scattering intensity thanks to the parameter ϕ. An empirically viable approximation for F is the term $R_s^{1.5}$. This was also used in the case of the milk sample.

Equation (12.2) was also used in an experiment in which the refractive indices and fat concentrations of various types of milk were studied using a reflectometer. Since the fat globules in milk scatter light there is reason to assume that the fat concentration and the scattering parameter ϕ exist in a mutually dependent and useful relationship. In this experiment the fat concentration of the milk was varied by diluting the milk with water and by

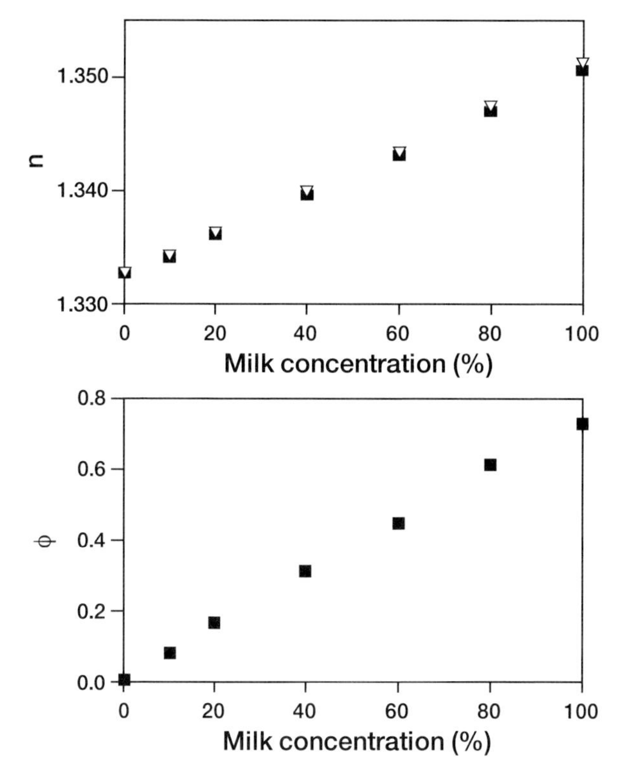

Fig. 12.6. The refractive index n and the scattering parameter ϕ of milk-water solutions. Reference measurement (Abbe refractometer) denoted by triangles

mixing milks with differing fat concentrations. The results are presented in Figs. 12.6 and 12.7. They show that both scattering and refractive index exist in linearly dependent relationships with fat concentration. It should be borne in mind that other components in milk besides fat globules also contribute to scattering. This is why fat-free milk does not have a scattering parameter of zero, as may be seen from the diagram 12.7.

12.2.2 Azzam's Polarization Method

The method developed by Azzam (Azzam's theory is presented in Sect. 4.2) is interesting in that its calculation requires only a minimal amount of measurement data, in fact, just two reflectances, in order to determine the optical constants. Thus the total measuring time may be short. The necessary data is obtained by measuring reflectance using s- and p-polarized light at a fixed angle of incidence and wavelength. The theory behind the method is elegant but requires accurate absolute reflectances in order to work (in this respect a calculation involving the R_s/R_p ratio would constitute a better alternative [229]). Furthermore, the choice of angle of incidence is of great relevance

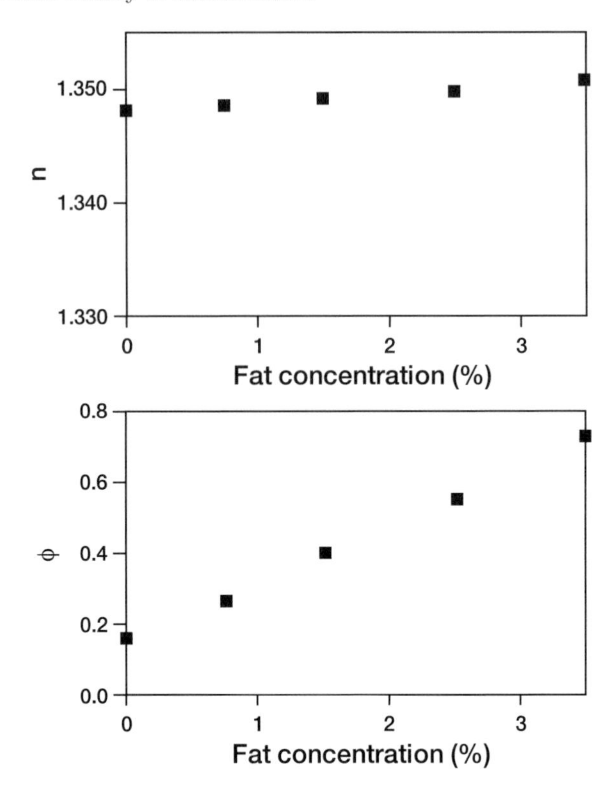

Fig. 12.7. The refractive index n and the scattering parameter ϕ determined at 590 nm for various milk samples

to the accuracy of the final result as was shown earlier on pages 36 and 37 of this book. The Azzam polarization method was tried out in the ethanol and lignin solutions discussed above. Since our interest here lay in the determination of refractive index the angle of incidence was set below the critical angle for both solutions. The results obtained were compared with those produced by a commercial refractometer (Bellingham & Stanley 60/ED). As seen in Figs. 12.8 and 12.9, the results obtained from the different methods are very similar.

The Azzam method is based on the calculation of the phase of reflection (4.12). Reflectance ($\theta < \theta_c$) from a non-absorbing material produces a phase shift δ equal to either zero or π, depending on the polarization of the incident light. Thus a non-zero phase shift is usually regarded as indicating the presence of absorption. However, it can be shown, as in [60], that under certain circumstances the phase shift δ may be zero even when the medium in question exhibits absorption characteristics. The conditions for this are that (a) the incident light be p-polarized and (b) the expression $\xi = 1/2|\epsilon_r|^2 Re(\epsilon_r)$ be such that $\xi \in [0, 1]$. Here ϵ_r is the relative dielectric function of the two media.

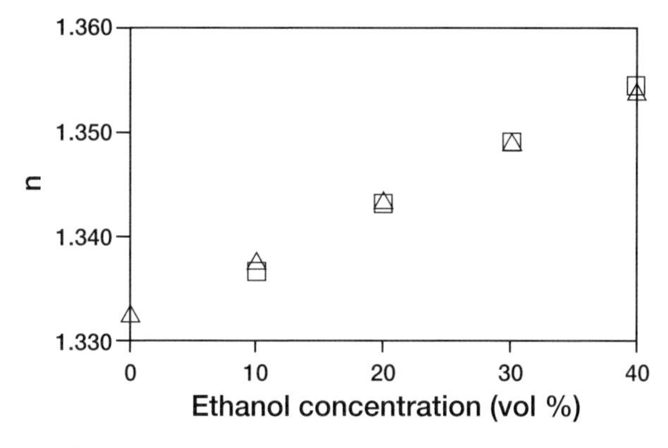

Fig. 12.8. The refractive indices of ethanol solutions determined using the Azzam method. Reference measurements were made employing an Abbe refractometer (triangles)

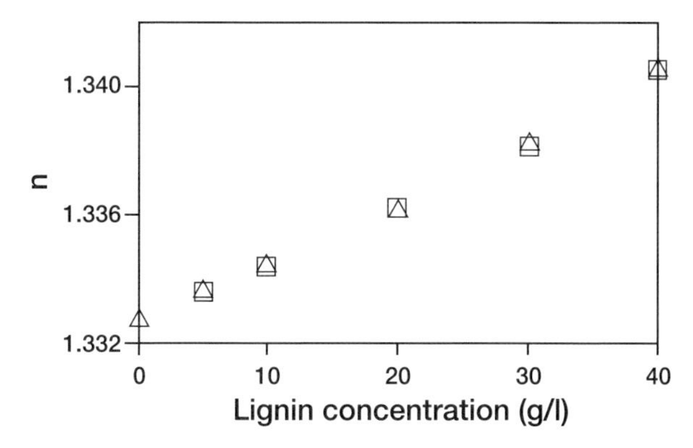

Fig. 12.9. The refractive indices of lignin solutions determined using the Azzam method. Reference measurements were made employing an Abbe refractometer (triangles)

The zero-phase then exists at the angle of incidence of $\theta_{\delta=0} = \arcsin(\xi^{1/2})$. For example, in Fig. 4.2 $\theta_{\delta=0} \approx 39°$.

12.2.3 SPR Method

The potential of surface plasmon resonance (SPR) for the characterisation of thin films and for monitoring processes at metal interfaces was recognised in the late seventies. Since then SPR-sensor technology has taken great strides forward both in terms of instrumentation development and applications. Much research work has been carried out within the areas of physics, chemistry and especially biology; commercial SPR sensors are nowadays available for the direct real-time observation of biomolecular interactions.

Surface plasmon can be excited in several ways. One of these, the Kretschmann configuration, exploits attenuated total reflectance. The other, most widely used configurations are the grating coupler-based SPR system and the optical waveguide-based SPR system. Experimental studies have been carried out using optical fibres, planar waveguides and prisms coated with single or multiple metal layers.

As well as studies within the field of biology, SPR sensors can be used to determine the optical constants of liquid phases. From the $R = R(\theta)$ curve given by a SPR sensor the real refractive index and the extinction coefficient can be evaluated. Typically, the angle of incidence where minimum reflectance occurs determines the real part of the complex refractive index while the shape of the reflectance curve is defined by the imaginary part of complex refractive index.

The reflectometer may be set in SPR measurement mode by exchanging the traditional prism for a prism which has one surface coated with a thin metal film. This film, therefore, separates the prism from the liquid sample. The surface plasmon resonance effect may be achieved by finding the angle of incidence which causes excitation in the angle range of total reflectance. At this particular angle of incidence the electrons in the metal begin to vibrate vigorously and, as a result, the energy of the incoming radiation is absorbed. This phenomenon occurs only with p-polarised light. Resonance is affected by the refractive indices of the materials and, therefore, also by wavelength. It should be noted that the reflectometer employed here also allows for the adjustment of wavelength in this measurement mode.

Typical SPR measurement mode $R = R(\theta)$ curves determined over a range of wavelengths (600–800 nm) are presented in Fig. 12.10. Data was obtained from a system consisting of a prism, a silver film ($\sim$ 50 nm in thickness) and water. From the diagram we note that as wavelength increases resonance is achieved at an increasingly small angle of incidence. In other words, R_{min} shifts in the direction of smaller angles of incident as wavelength increases. If we know the properties of the prism and the metal film we can use (4.51) to calculate the refractive index of an unknown sample. In practice, the accuracy of the measurement depends on how well we know the permittivity of the metal film.

The dip in the reflectance curve may be made very steep by employing a suitable experimental configuration, i.e. by optimising the thickness of the metal film. This makes the instrument a sensitive meter of refractive index. If we set the angle of incidence within the region of steep gradient either side of the centre of the dip then even a small change in the refractive index in a liquid will cause a measureable change in reflectance. This idea was tested in an experiment, the results of which are presented in Fig. 12.11. In the experiment pure ethanol was added in small (1 ml) quantities at fixed intervals of time to a quantity (initially 500 ml) of water. The reflectance of the solution at a wavelength of 670 nm was monitored throughout (in fact

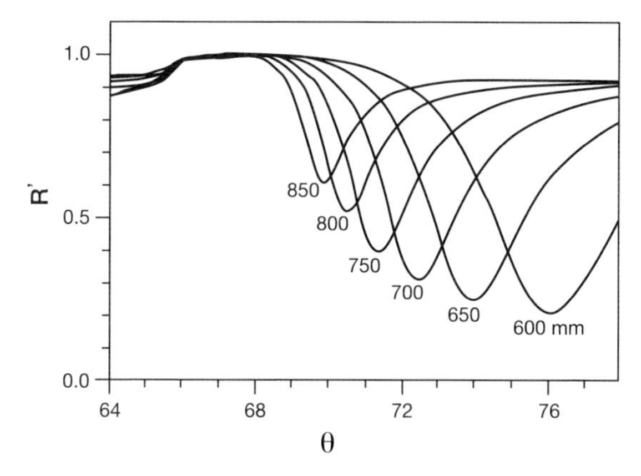

Fig. 12.10. The SPR signal of water at various wavelengths. The dips are a consequence of plasmon resonance

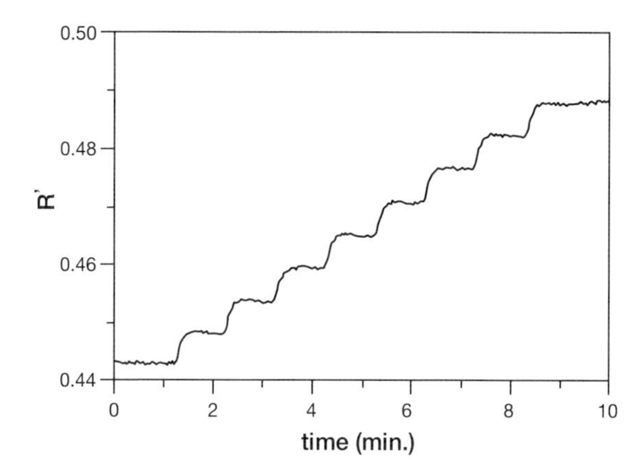

Fig. 12.11. The SPR signal as a function of time. The refractive index of the liquid was increased by adding ethanol into water. Here, the angle of incidence was set in the area of steepest gradient of the $R'(\theta)$-curve

using the reflectometer presented in Figs. 12.1 and 12.2 we monitored the voltage ratio R' of Detector 2 and Detector 1, which is, of course, closely related to the true reflectance). Having the higher refractive index of the two liquids, the addition of ethanol tends to increase overall refractive index of the solution. In the above case each additional 1 ml quantity of ethanol brought about an increase in refractive index of 0.00009 refractive index units (RIU). The clear stepped profile indicates that the experimental set-up would have been capable of detecting still smaller changes in the density of the solution. In addition, it is worth mentioning that no special procedures were carried out before the measurements which would have caused a steepening of the dip.

There are many ways of making use of wavelength adjustability in SPR measurements. It may be used to optimise measurement or even to further provoke the effect, especially if we have at our disposal only a small selection of prism-metal combinations. In addition, wavelength adjustment is an unconditional requirement if we are studying the absorption behaviour of a sample or the wavelength dependence of its refractive index.

12.2.4 Maximum Entropy Model, MEM

The MEM method presented in Sect. 6.4 provides information concerning the phase of amplitude reflection. The data required by the method consists of the reflectance curve as a function of wavelength together with one or more anchor points. The anchor points may be the refractive index or extinction coefficient or both determined for a particular wavelength range. When the phase data is calculated by the MEM method the complex refractive index of the sample may be calculated using Fresnel's equations. This produces $n(\lambda)$ and $k(\lambda)$ data about the sample.

We tested this phase retrieval method on a number of test samples. In one experiment we prepared test solutions from red food colouring and synthetic lignin. The red dye solution was measure over the 400–600 nm range with the angle of incidence fixed at 65.0 degrees while the lignin solution was measured over 250–450 nm using the same angle of incidence. In both cases the reflectance spectra were measured at wavelength intervals of 2 nm. The reflectance curves obtained are shown in Figs. 12.12a and 12.13a. The anchor points could, in principle, be freely chosen from the wavelength range used. In this experiment we chose two anchor points for each sample, i.e. the refractive indices determined at 450 nm and 525 nm for the red dye, and at 280 nm and 400 nm for lignin. The refractive indices for the anchor points were calculated using Fresnel's fitting method. Using this data the MEM calculation produces the results presented in Figs. 12.12b and c, 12.13b and c.

The same test solutions were also studied using a traditional refractometer (Bellingham & Stanley 60/ED), a spectrophotometer (Hitachi U-3300) for performing transmission measurements and, finally, by the ATR method. The results obtained from the MEM method match closely with these reference results. We note that for the transmission measurements the samples were diluted in water by factors of $(1:125)$ for the red dye and $(1:1333)$ for the lignin and placed in a 10 mm cuvette. In the refractometer we used two light sources, i.e. a mercury lamp and a sodium lamp. These produce narrow emission lines at 435.8 nm and 589.6 nm, respectively. ATR measurement was performed using the reflectometer by setting the angle of incidence above the critical angle. Although ATR measurement produces an absorption spectrum which closely resembles the spectra produced by transmittance measurements there are, nevertheless, fundamental differences between the respective mechanisms of generating spectra.

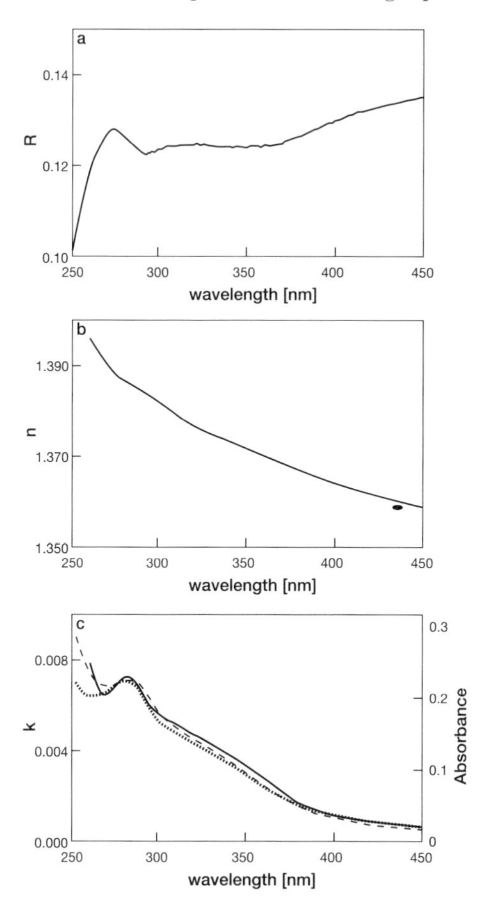

Fig. 12.12. (**a**) The experimental reflectance curve for a lignin solution. (**b**) The refractive index and (**c**) extinction coefficient curves calculated using the MEM method. The reference measurements are denoted as follows: dot – Abbe refractometer, dotted line – transmission and dashed line – ATR method (scale on the right side of the graph)

MEM is a quick and relatively accurate method for determining the wavelength dependence of optical constants. One particular advantage of this method is that it does not involve data extrapolation as does the KK method.

12.2.5 Attenuated Total Reflection ATR

In Sects. 4.4 and 4.5 of the first part of this book we dealt with the theory of total reflection. Light experiences total reflection at the boundary between two materials when the angle of incidence of the light beam exceeds the critical angle. The condition for the generation of total reflection concerns the values of the refractive indices of the two materials. Thus, total reflection

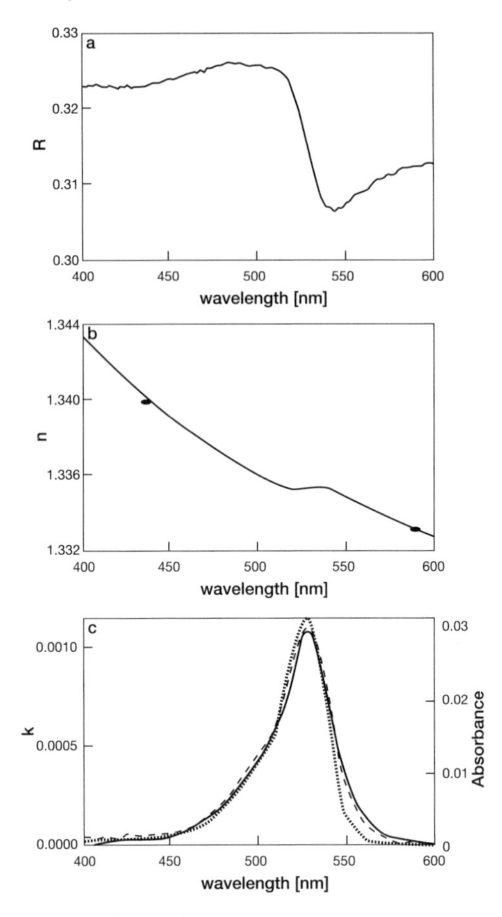

Fig. 12.13. (**a**) The experimental reflectance curve of a red colour solution. (**b**) The refractive index and (**c**) extinction coefficient curves calculated using the MEM method. Reference measurements are denoted as follows: dots – Abbe refractometer, dotted line – transmission and dashed line – ATR method (scale on the right side of the graph)

occurs when $n_1 > n_2$. Here the material through which the light travels on the way to the boundary and through which it continues to travel after reflection is designated by the number 1. Correspondingly, the material lying on the other side of the boundary is referred to by the number 2.

Total reflection generates in material 2 an evanescent wave which is attenuated exponentially as it propagates through the material. Thus, the material 2 may absorb the energy of the incident light via the evanescent wave. Reflection is attenuated and reflectance takes a value of less than one. This effect is known as attenuated total reflection, ATR.

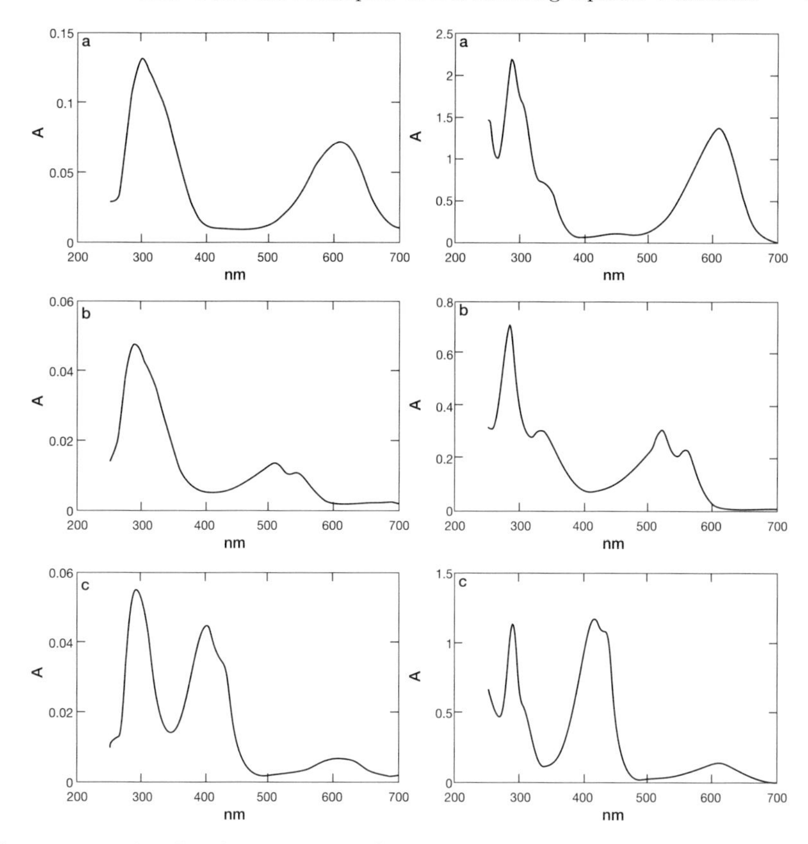

Fig. 12.14. The absorbance curves of blue, red and green dye solutions using the ATR mode of the reflectometer (left side) and a transmission method (right side). The liquids employed in a transmission measurement were diluted (1 : 100)

The ATR method may be exploited in the investigation of absorption spectra. To perform this we set the angle of incidence within the range of total reflection and then scan through the chosen wavelength range. It should, however, be borne in mind that there is one essential difference between transmission and ATR measurements. In transmission measurements the electric field can vibrate only in a plane perpendicular to the direction of the light path whereas in ATR measurement the electric field has components in all three dimensions. This in part may explain the differences in the spectra produced by the two methods. Figure 12.14 presents the absorption spectra of three different dye solutions measured using both ATR and transmission methods.

The sensitivity of ATR measurement may be improved by increasing the number of reflections. As only one reflection occurs in our reflectometer we should not expect the device to be able to detect very low concentrations. In the commercial UV/visible range ATR probe presented in Fig. 12.15 light is

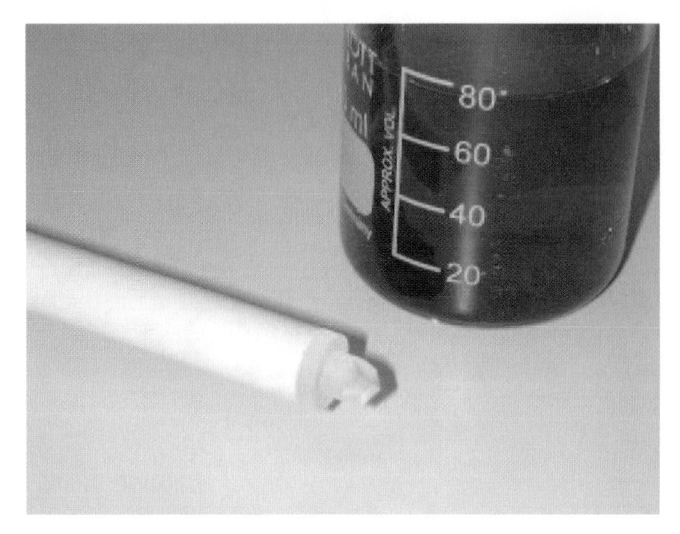

Fig. 12.15. A photograph of a commercial 3-reflection ATR probe

directed through a rod to a prism in which it is reflected three times before returning for analysis.

ATR measurements are widely used in the IR range. Additional devices based on the same method are available which may be fitted to the sample chamber of a spectrophotometer. The ATR method is particularly suitable when rapid spectra measurements are required. As the information comes from a very thin layer on the probe surface the method may suffer from contamination.

12.2.6 Measurement of Pigment Slurries

As an example of the utilization of the reflectometer in assessment of the refractive index of spherical pigment particles we consider here weakly absorbing styrene-acryl copolymer pigment spheres with concentric spherical shells [42]. The pigments are in water matrix and the volume concentration of the pigments is considerable high so that the binary pigment-water system is a slurry. A valuable tool in the investigation of the geometry of the pigments is a transmission electron microscope (TEM). In Fig. 12.16 is shown a TEM micrograph of an array of plastic pigments (PP). The interesting feature with present pigments is that there is water inside the core. Obviously the pigments can be approximated to take a spherical shape. As an average the diameter D of the pigment was 400 nm in this particular case.

In Fig. 12.17 are shown reflectance curves of a plastic pigment slurry. Curves were measured as function of the angle of incidence using s-polarized light and for three fixed wavelengths namely 400, 500 and 600 nm.

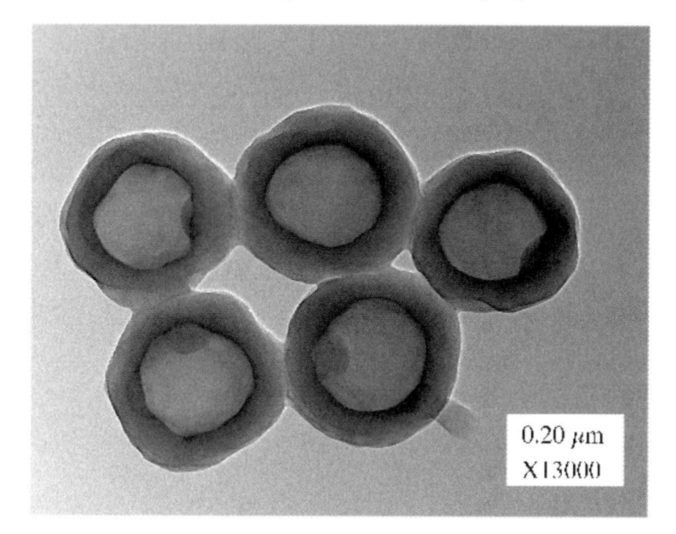

Fig. 12.16. TEM micrograph of an array of plastic pigments [42]

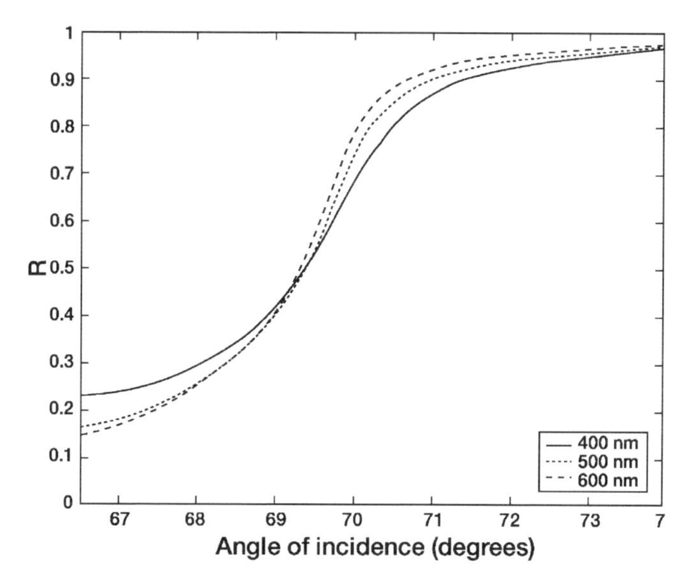

Fig. 12.17. Reflection from plastic pigment slurry as a function of light incidence. Solid line, 400 nm; dotted line, 500 nm; dashed line, 600 nm

First we calculated the effective refractive index of the slurry with the aid of (12.1). Next, since the refractive index of the water is a priori well known at visible range, we solved the refractive index of the spherical pigment using (3.59) and (3.60). The results of the measurements and calculations are shown in Table 12.1.

Table 12.1. Refractive index of the water and effective refractive index of the slurry and a PP particle at three wavelengths (D=400 nm)

λ (nm)	n_{water}	$n_{\text{slurry,reflectometer}}$	n_{particle}
400	1.3431	1.3763	1.5316
500	1.3363	1.3708	1.5227
600	1.3325	1.3668	1.5129

Table 12.2. Wiener bounds for the plastic-pigment slurry (D=400 nm)

λ (nm)	$n_{\text{slurry,reflectometer}}$	$n_{\text{slurry,lower bound}}$	$n_{\text{slurry,upper bound}}$
400	1.3763	1.3752	1.3829
500	1.3708	1.3681	1.3756
600	1.3668	1.3634	1.3705

Table 12.3. Wiener bounds for the plastic material (D=400 nm)

λ (nm)	$n_{\text{shell,lower bound}}$	$n_{\text{shell,upper bound}}$
400	1.5779	1.5967
500	1.5685	1.5869
600	1.5573	1.5746

Next the Wiener inequalities (3.54) were exploited for the calculation of the lower and upper bounds of the slurry using the effective refractive index of the spherical particle presented in Table 12.1. The results of the calculations are shown in Table 12.2. As can be seen fairly tight limits were obtained if compared to the measured refractive index of the slurry. Finally the intrinsic refractive index of the plastic shell was estimated using (3.55). The results of the calculations are presented in Table 12.3.

The above data of refractive index was used as anchor points in order to obtain the wavelength dependent refractive index and attenuation coefficient of he plastic pigment with the aid of MEM [230].

12.3 Contamination of the Probe

Some liquids react with the prism or other types of measurement window in such a way as to form a layer of material, the optical properties of which differ from those of the liquid. This phenomenon may be result from chemical activity or mechanical binding. Since reflection measurement produces

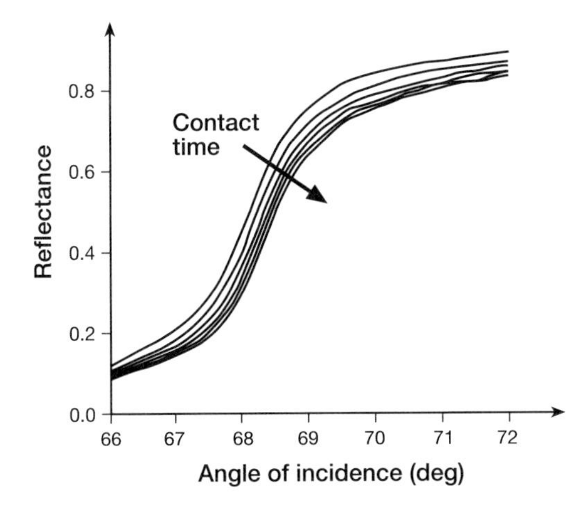

Fig. 12.18. The generation of a contaminant layer may be observed from a decreasing reflectance curve

information from only a thin layer at the boundary of the sample it is not surprising that interpretation of the measurement signal becomes difficult in such cases and that the results obtained do not necessarily describe the true properties of the liquid. The situation is not so serious in the case of transmission measurements where the thickness of the liquid layer under investigation is considerably greater than in reflection measurements. In practice, this problem of contamination is usually solved simply by cleaning the probe sufficiently often.

The "contamination layer" formed on the prism renders invalid the supposition of the homogeneity of the materials as required by Fresnel's equations. In terms of optics the contamination layer makes the refractive index and extinction coefficient dependent on the distance from the prism. We may thus speak of refractive index and extinction coefficient profiles. It is reasonable to expect that the material density of the contamination layer is greater than that of the liquid behind it. This assumption is borne out by the study of contaminating liquids using the reflectometer. It has been shown that at constant wavelength and angle of incidence reflectance falls off with time when the measurement prism comes into contact with a contaminating flowing liquid. When considered in the light of Fresnel's equations this decrease in reflectance is a direct consequence of the increase in refractive index. Figure 12.18 describes the effects of a contaminating lignin solution on reflectance, i.e. on the $R = R(\theta)$ curve. The solution was circulated through the system for a total of five hours during which time the R curves were measured at half hourly periods. The wavelength of the light used was 300 nm and the incident beam was p-polarised. The same effect is also observed for s-polarised light.

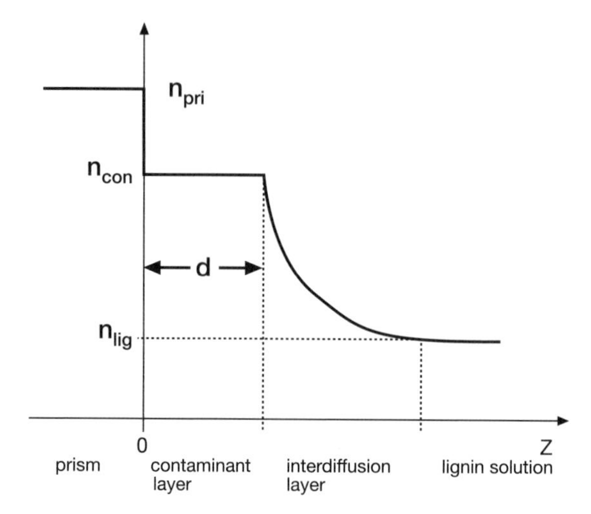

Fig. 12.19. A refractive index (extinction coefficient) profile for the study of a contaminant layer. Here n_{pri}, n_{con} and n_{lig} refer to refractive indices of prism, contaminant layer and lignin solution, respectively

The contamination layer causes the generation of refractive index and extinction coefficient profiles, as a consequence of which the fitting of Fresnel's equations for a single boundary no longer gives satisfactory results. We may, however, divide the profile into a number of sub-layers and suppose that the refractive index and extinction coefficient of each sub-layer remain constant. This then allows the fitting based on Fresnel's equations to the system. Reflection from a layered material has been described in numerous sources. In our study we used the matrix method presented in Sect. 5.3 of this book. Thus, provided we know the profile then we may calculate the reflectance from the equations of this matrix method. However, problems arise if the matter is considered in the opposite order. We do not know with certainty the form of the profile and in practice the measurement data is not sufficiently accurate for determining the profile. This problem and its solutions are discussed in references [111–116].

In the case of the above-mentioned lignin solution we obtained a good fit between the layer theory employed and the measurement data by supposing that the refractive index and extinction coefficient of the contamination layer are constant throughout the entire layer. Furthermore, we assumed that between the contamination layer and the lignin solution is an interdiffusion layer, the refractive index and extinction coefficient of which falls exponentially from the values in the contamination layer to those in the lignin solution (see Fig. 12.19). Thus the n- and k-profiles may be expressed as follows

$$\begin{aligned} n(x) &= a_n e^{-\frac{1}{2}b_n\{(x-d)+|x-d|\}} + n_{\mathrm{lig}} \\ k(x) &= a_k e^{-\frac{1}{2}b_k\{(x-d)+|x-d|\}} + k_{\mathrm{lig}}\,. \end{aligned} \qquad (12.3)$$

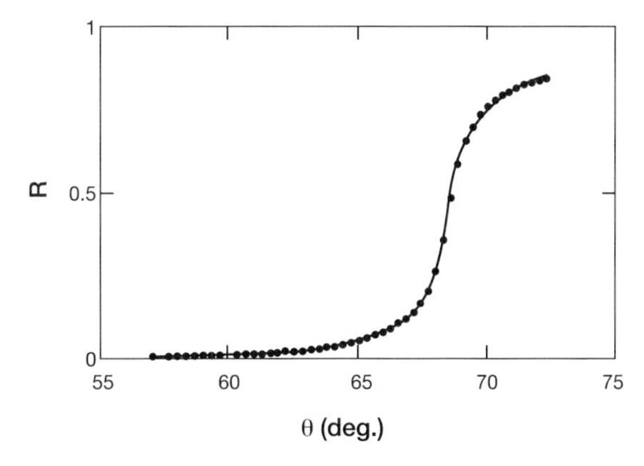

Fig. 12.20. The fitting result for data collected during 3.5 h contact time for a process liquid and a probe. Calculated reflectance and experimental data are represented by a solid line and dots, respectively

Here a_n, b_n, a_k and b_k are the coefficients of the n- and k-profiles, d is the thickness of the contaminant layer and x is the distance from the prism surface. The refractive index and the extinction coefficient of lignin solution is denoted as n_{lig} and k_{lig}, respectively.

When the measurement data is fitted to the theory explained above, as has been performed in Fig. 12.20, the optical constants of the profile may be determined. Further to this the calculation tells us the thickness of the contamination layer – in the case of the above diagram it was approximately 160 nm [231].

A Intensity Law for Nonlinear Absorption

We can write (3.32) in the form

$$\frac{dI}{I[\alpha + \gamma I]} = -dx \ . \tag{A.1}$$

Next we integrate the differential equation (A.1) using partial fractions

$$\frac{AdI}{I} + \frac{BdI}{\alpha + \gamma I} = -dx \ . \tag{A.2}$$

The coefficients A and B fulfill the following relations

$$\begin{aligned} A\gamma + B &= 0 \\ A\alpha &= 1 \ . \end{aligned} \tag{A.3}$$

Thus $A = 1/\alpha$ and $B = -\gamma/\alpha$. The integration of (A. 2) yields now

$$\frac{1}{\alpha}\int_{I_0}^{I} \frac{dI}{I} - \frac{1}{\alpha}\int_{I_0}^{I} \frac{dI}{I + \frac{\alpha}{\gamma}} = -\int_0^x dx \ . \tag{A.4}$$

Thus we get a solution, which is a sum of two logarithmic functions

$$\frac{1}{\alpha}\ln\frac{I}{I_0} - \frac{1}{\alpha}\ln\left(\frac{I + \frac{\alpha}{\gamma}}{I_0 + \frac{\alpha}{\gamma}}\right) = -x \ . \tag{A.5}$$

The exponent law of (3.32) is obtained from (A.5) using the well-known addition properties of logarithm and some algebra as follows:

$$\frac{1}{\alpha}\ln\left(\frac{I}{I_0}\frac{I_0 + \frac{\alpha}{\gamma}}{I + \frac{\alpha}{\gamma}}\right) = -x \ , \tag{A.6}$$

therefore it holds that

$$\frac{I + \frac{\alpha}{\gamma}}{I} = \frac{I_0 + \frac{\alpha}{\gamma}}{I_0} e^{\alpha x} \ . \tag{A.7}$$

Equation (3.33) follows from (A.7).

B Complex Angle of Refraction

In mathematical sense it is possible to substitute a complex angle of refraction, $\Psi = \xi + \mathrm{i}\zeta$ in (4.28) and (4.29). Then we can require that

$$\begin{aligned} \cos\Psi &= \cos(\xi + \mathrm{i}\zeta) = \mathrm{i}C \\ \sin\Psi &= \sin(\xi + \mathrm{i}\zeta) = D > 1\,, \end{aligned} \tag{B.1}$$

where C and D are real constants. According to the Euler's formula it holds that

$$e^{\mathrm{i}\Psi} = \cos\Psi + \mathrm{i}\sin\Psi = \mathrm{i}C + \mathrm{i}D = \mathrm{i}(C + D)\,. \tag{B.2}$$

We can also write

$$e^{\mathrm{i}z} = e^{\mathrm{i}(x+\mathrm{i}y)} = e^{-y}e^{\mathrm{i}x}\,. \tag{B.3}$$

Thus the modulus of (B.3) is

$$|e^{-y}e^{\mathrm{i}x}| = |\mathrm{i}(C + D)| = C + D\,. \tag{B.4}$$

From (B.4) we can solve the equation

$$e^{-y} = C + D\,, \tag{B.5}$$

which means that

$$y = \ln\frac{1}{C + D}\,. \tag{B.6}$$

If we choose $x = \pi/2$ and apply the identity $\mathrm{i} = e^{\mathrm{i}\pi/2}$ in (B.2) we find out finally the complex angle

$$\Psi = \frac{\pi}{2} + \mathrm{i}\ln\frac{1}{C + D}\,. \tag{B.7}$$

C Cauchy–Riemann Equations

The analyticity of complex relative permittivity $\epsilon = \epsilon(\Omega)$ can be tested using Cauchy-Riemann equations. If the relative permittivity is an analytic function it holds that

$$\begin{aligned} \frac{\partial Re\{\epsilon_r(\omega,\xi)\}}{\partial\omega} &= \frac{\partial Im\{\epsilon_r(\omega,\xi)\}}{\partial\xi} \\ \frac{\partial Re\{\epsilon_r(\omega,\xi)\}}{\partial\xi} &= -\frac{\partial Im\{\epsilon_r(\omega,\xi)\}}{\partial\omega} \ . \end{aligned} \tag{C.1}$$

In the case the relative permittivity is a simple function such as that of Lorentz (6.1) one can directly form the complex derivative

$$\frac{d\epsilon_r(\Omega)}{d\Omega} = \frac{\rho e^2}{m\epsilon_0} \frac{2\Omega + \mathrm{i}\Gamma}{(\omega_0^2 - \Omega^2 - \mathrm{i}\Gamma\Omega)^2} \ , \tag{C.2}$$

which by inspection is observed to be finite and continuous function everywhere except at the poles that are located in the lower half of complex angular frequency plane.

D Kramers–Kronig Integrals and Symmetry Properties

The integral

$$P\int_{-\infty}^{\infty}\frac{N(\omega)-1}{\omega-\omega'}d\omega = \mathrm{i}\pi[N(\omega')-1] \tag{D.1}$$

can be separated to real and imaginary parts as follows:

$$\begin{aligned} n(\omega')-1 &= \frac{1}{\pi}P\int_{-\infty}^{\infty}\frac{k(\omega)}{\omega-\omega'}d\omega \\ k(\omega') &= -\frac{1}{\pi}P\int_{-\infty}^{\infty}\frac{n(\omega)-1}{\omega-\omega'}d\omega \ , \end{aligned} \tag{D.2}$$

which usually are denoted as a Hilbert transform pair. The expressions of (D.2) are not practical in physics since they involve integration along the negative angular frequency axis. Fortunately, the complex refractive index obeys the symmetry relations (6.19). Using the information of the symmetry relations (6.19) we get rid of the integration along the negative axis and obtain integrals i.e. K–K relations, which involve physically reasonable integration only on the positive axis. As an example we derive the K–K relation related to the first Hilbert transform in (D.2). It holds that

$$\begin{aligned} n(\omega')-1 &= \frac{1}{\pi}P\int_{0}^{\infty}\frac{k(\omega)}{\omega-\omega'}d\omega - \frac{1}{\pi}P\int_{0}^{-\infty}\frac{k(\omega)}{\omega-\omega'}d\omega \\ &= \frac{1}{\pi}P\int_{0}^{\infty}\frac{k(\omega)}{\omega-\omega'}d\omega - \frac{1}{\pi}P\int_{0}^{\infty}\frac{k(-\omega)}{-\omega-\omega'}(-d\omega) \\ &= \frac{2}{\pi}P\int_{0}^{\infty}\frac{\omega k(\omega)}{\omega^2-\omega'^2}d\omega \ . \end{aligned} \tag{D.3}$$

The derivation of the other K–K relation is analogous to the derivation above.

References

1. J.R. Partington: *Advanced Treatise on Physical Chemistry* vol 4. (Longmans, Norwich 1960)
2. L.A. Vanderberg: Appl. Spectrosc. **54**, 376A (2000)
3. R.J. Green, R.A. Frazier, K.M. Shakesheff, M.C. Davies, C.J. Roberts, S.J.B. Tendler: Biomaterials **21**, 1823 (2000)
4. S.G. Warren: Appl. Opt. **23**, 1206 (1984)
5. P. Schiebener, J. Straub, J.M.H.L. Sengers, J.S. Gallagher: J. Phys. Ch R **19**, 677 (1990)
6. R.A.J. Litjens, T.I. Quickenden, C.G. Freeman: Appl. Opt. **38**, 1216 (1999)
7. R. Briggs, K.T.V. Grattan: Trans. Inst. MC **12**, 65 (1990)
8. C.F. Bohren, D.R. Huffman: *Absorption and Scattering of Light by Small Particles* (Wiley, New York 1983)
9. F. Wooten: *Optical Properties of Solids* (Academic Press, New York 1972)
10. C. Kittel: *Introduction to Solid State Physics*, 3rd edn. (Wiley, New York 1968)
11. M. Born, E. Wolf: *Principles of Optics*, 6th edn. (Pergamon, New York 1980)
12. S.S. Batsanov: *Refractometry and Chemical Structure* (Van Nostrand, New Jersey 1966)
13. K.-E. Peiponen, E.M. Vartiainen, T. Asakura: *Dispersion, Complex Analysis and Optical Spectroscopy* (Springer, Heidelberg 1999)
14. R.L. Sutherland: *Handbook of Nonlinear Optics* (Marcel Dekker, New York 1996)
15. Y.R. Shen: *The Principles of Nonlinear Optics* (John Wiley and Sons, New York 1984)
16. A. Dragomir, J.G. McInerney, D.N. Nikogosyan, A.A. Ruth: IEEE J. Quant. Electron **38**, 31 (2002)
17. G.M. Holdridge: *Nanostructure Science and Technology R and D Status and Trends in Nanoparticles, Nanostructured Materials, and Nanodevices*(Kluwer, Dordrecht 1999)
18. M.N.V. Ravi Kumar: J. Pharm. Pharmaceut. Sci. **3**, 234 (2000)
19. P. Hänninen, J.T. Soini, E. Soini: Cytometry **36**, 183 (1999)
20. P.E. Hänninen, A. Soini, N. Meltola, J. Soini, J. Soukka, E. Soini: Nature Biotechnology **18**, 538 (2000)
21. J.C. Maxwell Garnett: Trans. R. Soc. **203**, 385 (1904)
22. J.C. Maxwell Garnett: Trans. R. Soc. **205**, 237 (1906)
23. D.A.G. Bruggeman: Ann. Phys. **24**, 636 (1935)
24. R.W. Boyd, R.J. Gehr, G.L. Fischer, J.E. Sipe: Pure Appl. Opt. **5**, 505 (1996)
25. J.D. Jackson: *Classical Electrodynamics* (Wiley, New York 1962)
26. D.E. Aspnes: Am. J. Phys. **50**, 704 (1982)

27. K.-E. Peiponen, M.O.A. Mäkinen, J.J Saarinen, T. Asakura: Opt. Rev. **8**, 9 (2001)
28. R. Ruppin: Opt. Commun. **182**, 273 (2000)
29. D.E. Aspnes, J.B. Theeten, F. Hottier: Phys. Rev. B **20**, 3292 (1979)
30. I.L. Skryabin, A.V. Radchik, P. Moses, G.B. Smith: Appl. Phys. Lett. **70**, 2221 (1997)
31. G.L. Fischer, R.W. Boyd, R.J. Gehr, S.A. Jenehke, J.A. Osaheni, J.E. Sipe, L.A. Weller-Brophy: Phys. Rev. Lett. **74**, 1871 (1995)
32. D. Faccio, P. Di Trapani, E. Borsella, F. Gonella, P. Mazzoldi, A.M. Malvezzi: Europhys. Lett. **43**, 213 (1998)
33. X.C. Zeng, D.J. Bergman, P.M. Hui, D. Stroud: Phys. Rev. B **38**,10970 (1988)
34. A.E. Neeves, M.H. Birnboim: J. Opt. Soc. Am. B **6**, 787 (1989)
35. J.W. Haus, H.S. Zhous, S. Takami, M. Hirasawa, I. Honma, H. Komiyama: J. Appl. Phys. **73**,1043 (1993)
36. D.D. Smith, G. Fischer, R.W. Boyd, D.A. Gregory: J. Opt. Soc. Am. B **14**, 1625 (1997)
37. J.E. Sipe, R.W. Boyd: Phys. Rev. A , 1614 (1992)
38. K.-E. Peiponen, E.M. Vartiainen, T. Asakura: J. Phys. Condens. Matter **10**, 2483 (1998)
39. R.J. Gehr, G.L. Fischer, R.W. Boyd: J. Opt. Soc. Am. B **14**, 2310 (1997)
40. K.-E. Peiponen, E.M. Vartiainen, J.J. Saarinen, M.O.A. Mäkinen: Opt Commun **205**, 17 (2002)
41. O. Wiener: Abh. Math. Phys. K1 Königl. Ges. **32**, 509 (1912)
42. A.J. Jääskeläinen, K.-E. Peiponen, J. Räty, U. Tapper, O. Richard, E.I Kauppinen, K. Lumme: Opt. Eng. **39**, 2959 (2000)
43. H.C. Van de Hulst: *Light Scattering by Small Particles* (Wiley, New York 1957)
44. M.Kerker: *The Scattering of Light* (Academic, New York 1969)
45. V.P. Maltsev, A.V. Chernyshev, K.A. Sem'yanov, E. Soini: Meas. Sci. Technol. **8**, 1023 (1997)
46. A. Killey, G.H. Meeten: J. Chem. Soc. Faraday Trans. **77**, 587 (1981)
47. G.H. Meeten, A.N. North: Meas. Sci. Technol. **6**, 214 (1995)
48. G.H. Meeten: Opt. Commun. **134**, 233 (1997)
49. M. Mohammadi: Advances in Colloidal and Interface Science **65**, 17 (1995)
50. J.C. Dijt, M.A. Cohen Stuart, G.J. Fleer: Advances in Colloidal and Interface Science **50**, 79 (1994)
51. G. Kortüm: *Reflectance Spectroscopy Principles, Methods, Applications* (Springer, Berlin 1969)
52. N.J. Harrick: *Internal Reflection Spectroscopy* (Harrick Scientific Corporation, New York 1979)
53. F.M. Mirabella (Ed.): *Internal Reflection Spectroscopy: Theory and Applications* (Dekker, New York 1993)
54. R.T. Holm: *Conventions Confusions. In: E. W. Palik (Ed.) Handbook of Optical Constants of Solids II* (Academic Press, Orlando 1998)
55. S.P.F. Humphreys-Owen: Proc. Phys. Soc. **77**, 949 (1961)
56. C. Croxton: *Introduction to Liquid State Physics* (John Wiley and Sons, London 1975)
57. M. Saito, N. Matsumoto, J. Nishimura: Appl. Opt. **37**, 5169 (1998)
58. R.M.A. Azzam: J. Opt. Soc. Am. **68**, 1613 (1978)
59. R.M.A. Azzam: J. Opt. Soc. Am. **69**, 1007 (1979)

60. R.M.A. Azzam: J. Opt. Soc. Am. **69**, 487 (1979)
61. R.M.A. Azzam: Opt. Acta. **26**, 113 (1979)
62. R.M.A. Azzam: J. Opt. Soc. Am. **69**, 590 (1979)
63. J. Räty, K.-E. Peiponen: Meas. Sci. Technol. **11**, 74 (2000)
64. W.R. Hunter: *Measurement of Optical Constants in the Vacuum Ultraviolet Spectral Region. In: E. D. Palik (Ed.) Handbook of Optical Constnts of Solids* (Academic Press, London 1998)
65. G.H. Meeten: Meas. Sci. Technol. **8**, 728 (1997)
66. N.J. Harrick: J. Opt. Soc. Am. **55**, 851 (1965)
67. M.C. Peña-Gomar, A. Garcia-Valenzuela: Appl. Opt. **39**, 5131 (2000)
68. Q.W. Song, C.-K. Ku, C. Zhang, R.B. Gross, R.R. Birge, R. Michalak: J. Opt. Soc. Am. B **12**, 797 (1995)
69. S. Ekgasit: Appl. Spectrosc. **54**, 756 (2000)
70. MathSoft Inc: *Mathcad 6.0 User's Guide* (1995)
71. J. Räty, E. Keränen, K.-E. Peiponen: Meas. Sci. Technol. **9**, 95 (1998)
72. A.E. Kaplan: Sov. Phys. JETP **45**, 896 (1977)
73. P.W. Smith, J.-P. Hermann, W.J. Tomlinson, P.J. Maloney: Appl. Phys. Lett. **35**, 846 (1979)
74. K.-E. Peiponen, A. Jääskeläinen, J. Räty, O. Richard, U. Tapper, E.I. Kauppinen, K. Lumme: Appl. Spectrosc. **54**, 878 (2000)
75. J. Räty, K.-E. Peiponen: Appl. Spectrosc. **53**, 1123 (1999)
76. O. Kretschmann, H. Räther: Z Naturforsch. A **23**, 2135 (1968)
77. O. Kretschmann: Z Physik **241**, 313 (1971)
78. H. Räther: *Surface Plasmons on Smooth and Rough Surfaces and on Gratings* (Springer, Berlin 1988)
79. H. Räther: *Surface Plasma Oscillations and Their Applications. In: G. Hass, M. H. Francombe, and R. W. Hoffman (Eds) Physics of Thin Films vol 9* (Academic Press, New York 1977)
80. J. Homola, S.S. Yee, G. Gauglitz: Sens. Actuators B **54**, 3 (1999)
81. G.J. Sprokel, J.D. Swalen: *Attenuated Total Reflection Method. In: E. D. Palik (Ed) Handbook of Optical Constants of Solids II* (Academic Press, New York 1998)
82. A. Otto: Phys. Stat. Sol. **26**, K99 (1968)
83. A. Otto: Z Phys. **219**, 227 (1969)
84. R.M.A Azzam, N.M. Bashara: *Ellipsometry and Polarized Light* (North-Holland, Amsterdam 1977)
85. W.P. Chen, J.M. Chen: Surface Science **91**, 601 (1980)
86. W.P.Chen, J.M. Chen: J. Opt. Soc. Am. **71**, 189 (1981)
87. B.G. Tilkens, Y.F. Lion, Y.L. Renotte: Opt. Eng. **39**, 363 (2000)
88. H. Kano, S. Kawata: Appl. Opt. **33**, 5166 (1994)
89. C. Nylander, B. Liedberg, T. Lind: Sens. Actuators **3**, 79 (1982-183)
90. S. Löfås, M. Malmqvist, I. Rönnberg, E. Stenberg: Sensors and Actuators B **5**, 79 (1991)
91. K.-E. Peiponen, J. Räty, E.M. Vartiainen, T. Sugiura, S. Kawata: Meas. Sci. Technol. **10**, N145 (1999)
92. K.-E. Peiponen, A. Jääskeläinen, J. Räty, O. Richard, U. Tapper, E.I. Kauppinen, K. Lumme: Appl. Spectrosc. **54**, 878 (2000)
93. A. Jääskeläinen, K.-E. Peiponen, J. Räty: J. Dairy Sci. **84**, 38 (2001)
94. K. Matsubara, S. Kawata, S. Minami: Appl. Opt. **27**, 1160 (1988)

95. K. Johansen, R. Stålberg, I. Lundstöm, B. Liedberg: Meas. Sci. Technol. **11**, 1630 (2000)
96. L.M. Zhang, D. Uttamchandani: Electron Lett. **24**, 1469 (1988)
97. R.C. Jorgenson, C. Jung, S.S. Yee: Sens. Actuators B13 **13**, 721 (1993)
98. C.C. Jung, R.C. Jorgenson, C.H. Morgan, S.S. Yee: Process Control and Quality **7**, 167 (1995)
99. S.S. Karlsen, K.S Johnston, R.C. Jorgenson, S.S. Yee: Sens. Actuators B **24-25**, 747 (1995)
100. C.R. Lavers, J.S. Wilkinson: Sens. Actuators B **22**, 75 (1994)
101. J. Räty, K.-E. Peiponen, A. Jääskeläinen, M.O.A. Mäkinen: Appl. Spectrosc. **56**, 935 (2002)
102. J.J. Saarinen, K.-E. Peiponen, E.M. Vartiainen: Appl. Spectrosc. **57**, 288 (2003)
103. H.E. de Bruijn, R.P.H. Kooyman, J. Greve: Appl. Opt. **31**, 440 (1992)
104. H.-P. Chiang, Y.-C. Wang, P.T. Leung, W.S. Tse: Opt. Commun. **188**, 283 (2001)
105. B. Rothenhäusler, W. Knoll: Nature **332**, 615 (1988)
106. W. Hickel, B. Rothenhäusler, W. Knoll: J. Appl. Phys. **66**, 4832 (1989)
107. T. Okamoto, I. Yamaguchi: Opt. Commun. **93**, 265 (1992)
108. H. Knobloch, G. von Szada-Borryszkowski, S. Woigk, A. Helms, L. Brehmer: Appl. Phys. Lett. **69**, 2336 (1996)
109. H. Kano, W. Knoll: Opt. Commun. **182**, 11 (2000)
110. M.O.A. Mäkinen, K.-E. Peiponen, J. Räty, V. Hyvärinen: Appl. Spectrosc. **55**, 852 (2001)
111. L.J. Fina, G. Chen: Vib. Spectrosc. **1**, 353 (1991)
112. T. Hirschfeld: Appl. Spectrosc. **31**, 289 (1977)
113. H.G. Tompkins: Appl. Spectrosc. **28**, 335 (1974)
114. R.A. Shick, J.L. Koenig, H. Ishida: Appl. Spectros. **47**, 1237 (1993)
115. S. Ekgasit, H. Ishida: Appl. Spectrosc. **50**, 1187 (1996)
116. S. Ekgasit, H. Ishida: Appl. Spectrosc. **51**, 1488 (1997)
117. W. Weinstein: Computations in Thin Film Optics. In: *Vacuum* (E.T.Heron & Co. Ltd., London 1954)
118. A.Herpin: Comptes Rendus **225**, 182-183 (1947)
119. J.A. Dobrowolski: Optical Properties of Films and Coatings. In: *Handbook of Optics I* ed. by M. Bass (McGraw-Hill, New York 1995)
120. R. Kronig: J. Opt. Soc. Am. **12**, 547 (1926)
121. H.A. Kramers: Phys. Z **30**, 522 (1927)
122. R. Kronig: Ned. T. Natuurk **9**, 402 (1942)
123. J.S. Toll: Phys. Rev. **104**,1760 (1956)
124. H.M. Nussenzveig: *Causality and Dispersion Relations* (Academic Press, New York 1972)
125. L. Ahlfors: *Complex Analysis, 3rd edn.* (McGraw-Hill, New York 1979)
126. G.B. Arfken, H.J. Weber: *Mathematical Methods for Physicists, 4th edn.* (Academic Press, London 1995)
127. F. Bassani, M. Altarelli: *Interaction of Radiation with Condensend Matter. In: E. E. Koch (Ed) Handbook of Synchrotron Radiation, Vol 1(a)* (North-Holland, Amsterdam 1983)
128. D.E. Aspnes: *The Accurate Determination of Optical Properties by Ellipsometry. In: E. D. Palik (Ed) Handbook of Optical Constants of Solids* (Academic Press, Boston 1985)

129. K.-E. Peiponen, E.M. Vartiainen: Phys. Rev. B **44**, 8301 (1991)
130. T.G. Goplen, D.G. Cameron, R.N. Jones: Appl. Spetrosc. **34**,653 (1980)
131. M. Altarelli, D. Dexter, H.M. Nussenzveig, D.Y. Smith: Phys. Rev. B **6**, 4502 (1972)
132. M. Altarelli, D.Y. Smith: Phys. Rev. B **9**, 1290 (1974)
133. B. Velický: Chech. J. Phys. B **11**, 541 (1961)
134. D.Y. Smith: J. Opt. Soc. Am. **67**, 570 (1977)
135. R.K. Ahrenkiel: J. Opt. Soc. Am. **61**, 1651 (1971)
136. K.F. Palmer, M.Z. Williams, B.A. Budde: Appl. Opt. **37**, 2660 (1998)
137. F.W. King: J. Opt. Soc. Am. B **19**, 2427 (2002)
138. V.G. Foster: J. Phys. D: Appl. Phys. **25**, 525 (1992)
139. F.W. King: J. Chem. Phys. **71**,4726 (1979)
140. D.Y. Smith, C.A. Manogue: J. Opt. Soc. Am. **71**, 935 (1981)
141. M. Kogan : Sov. Phys. JETP **16**, 217 (1963)
142. P.J. Price: Phys. Rev. **130**, 1792 (1963)
143. W.J. Caspers: Phys. Rev. A , 1249 (1964)
144. F.L. Ridener, R.H. Good Jr: Phys. Rev. B **11**, 2768 (1975)
145. K.-E. Peiponen: Phys. Rev. B **37**, 6463 (1988)
146. H. Kishida, T. Hasegawa, Y. Iwasa, T. Koda: Phys. Rev. Lett. **70**, 3724 (1993)
147. F. Bassani, S. Scandolo: Phys. Rev. B **44**, 8446 (1991)
148. K.-E. Peiponen, E.M. Vartiainen, T. Asakura: *Dispersion Relations and Phase Retrieval in Optical Spectroscopy In: E. Wolf (Ed) Progress in Optics XXXVII* (North-Holland, Amsterdam 1997)
149. E. Tokunaga, A. Terasaki, T. Kobaysahi: Phys. Rev. A **47**, R4581 (1993)
150. E. Tokunaga, A. Terasaki, T. Kobayashi: J. Opt. Soc. Am. B **12**, 753 (1995)
151. E. Tokunaga, A. Terasaki, T. Kobayashi: J. Opt. Soc. Am. B **13**, 496 (1996)
152. S.G. Krantz: *Function Theory of Several Complex Variables* (Wiley, New York 1982)
153. R. Kronig: Ned. T Natuurk **9**, 402 (1942)
154. R. Nevanlinna: *Analytic Functions* (Springer, Berlin 1970)
155. F. Bassani, V. Lucarini: Eur. Phys. J. B **12**, 323 (1999)
156. K.-E. Peiponen: J. Phys. A: Math General (2001)
157. P.P Kircheva, G.B. Hadjichristov: J. Phys. B: Mol. Opt. Phys. **27**, 3781 (1994)
158. K.-E. Peiponen, J.J. Saarinen: Phys. Rev. A **65**, 063810-1 (2002)
159. K.-E. Peiponen: Opt. Rev. **4**, 433 (1997)
160. E.M. Vartiainen, K.-E. Peiponen, T. Asakura: Opt. Commun. **89**, 37 (1992)
161. J.P. Burg: Maximum entropy spectral analysis. In: textitProc. 37th Ann Meeting Soc Explor Geophysics, Oklahoma City (Oklahoma 1967)
162. J.K. Kauppinen, D.J. Moffatt, M.R. Hollberg, H.H. Mantsch: Appl. Spectrosc. **45**, 411 (1991)
163. J.K. Kauppinen, D.J. Moffatt, M.R. Hollberg, H.H. Mantsch: Appl. Spectrosc. **45**, 1516 (1991)
164. E.M. Vartiainen, K.-E. Peiponen, T. Asakura: Appl. Opt. **32**, 1126 (1993)
165. E.M. Vartiainen, T. Asakura, K.-E. Peiponen: Opt. Commun. **104**, 149 (1993)
166. E.M. Vartiainen, K.-E. Peiponen, T. Asakura: Appl. Spectrosc. **50**, 1283 (1996)
167. J. Räty, E.M. Vartiainen, K.-E. Peiponen: Appl. Spectrosc. **53**, 92 (1999)
168. E.M. Vartiainen: J. Opt. Soc. Am. B **9**, 1209 (1992)
169. E.M. Vartiainen, K.E. Peiponen, H. Kishida, T. Koda: J. Opt. Soc. Am. B **13**, 2106 (1996)

170. P.-K. Yang, J.Y. Huang: J. Opt. Soc. Am. B **14**, 2443 (1997)
171. K.-E. Peiponen, E.M. Vartiainen, T. Asakura: J. Phys.: Condens. Matter **9**, 8937 (1997)
172. K.-E. Peiponen, E.M. Vartiainen, T. Asakura: J. Phys.: Condens. Matter **10**, 2483 (1998)
173. S. Haykin, S. Kesler: Prediction-Error Filtering and Maximum Entropy Spectral Estimation. In: S Haykin (Ed) *Nonlinear Methods of Spectral Analysis* (Springer, Berlin 1983)
174. J.-F. Brun, D. De Sousa Meneses, B. Rousseau, P. Echegut: Appl. Spectrosc. **55**, 774 (2001)
175. E.M. Vartiainen, K.-E. Peiponen, T. Asakura: J. Phys: Condens. Matter **5**, L113 (1993)
176. H.-H. Perkampus: *UV-VIS Spectroscopy and Its Applications*(Springer-Verlag, Berlin 1992)
177. P. Cielo: *Optical Techniques for Industrial Inspection* (Academic Press, Boston 1988)
178. E.D. Zalewski: Radiometry and Photometry. In: *Handbook of Optics II* ed. by M. Bass (McGraw-Hill, New York 1995)
179. J.M. Palmer: The Measurement of Transmission, Absorption, Emission and Reflection. In: *Handbook of Optics II* ed. by M. Bass(McGraw-Hill, New York 1995)
180. W.R. McCluney: *Introduction to Radiometry and Photometry* (Artech House, Boston 1994)
181. E. Hecht: *Optics* 2nd edn.(Addison-Wesley, Massachusetts 1987)
182. K. Jokela: *Radiomerian perusteet ja Optisen säteilyn mittaukset*(Otatieto, Espoo 1992)
183. W.J. Smith: *Modern Optical Engineering*(McGraw-Hill, New York 1966)
184. R.W. Waynant, M.N. Ediger (Eds.): *Electro-Optics Handbook* (McGraw-Hill, New York 2000)
185. W.T. Walsh: *Photometry*, 3th edn. (Dover, New York 1965)
186. A. Stimson: *Photometry and Radiometry for Engineers* (Wiley, New York 1974)
187. D.A. Skoog, F.J. Holler, T.A. Nieman: *Principles of Instrumental analysis* 5th edn.(Saunders Collage Publishing, Philadelphia 1998)
188. F.E. Nicodemus: Appl. Opt. **9**, 1474 (1970)
189. D.B. Judd: J.Opt.Soc.Am. **57**, 445 (1967)
190. A. Rosencwaig: *Photoacoustics and photoacoustic spectroscopy*(Robert E. Krieger Publishing Company, Inc., Florida, 1990)
191. J.B. Birks: *Organic Molecular Photophysics* (Wiley, London 1973)
192. W. Elenbaas: *Light Sources* (MacMillan, Dorking 1772)
193. S. Svanberg: *Atomic and Molecular Spectroscopy*(Springer-Verlag, Berlin 1990)
194. A. LaRocca: Artificial Sources. In: *Handbook of Optics II* ed. by M. Bass(McGraw-Hill, New York 1995)
195. S. Miller: Noncoherent Sources. In: *Electro-Optics Handbook* ed. by R.W. Waynant, M.N. Ediger (McGraw-Hill, New York 2000)
196. Schott Glasswerke, Optics Division: *Optical Glass Catalogue* (Schott Glaswerke, Mainz 1984)
197. W.J. Tropf, M.E. Thomas, T.J. Harris: Properties of crystals and glasses. In: *Handbook of Optics II* ed. by M. Bass(McGraw-Hill, New York 1995)

198. J. Michl and E.W. Thulstrup: *Spectroscopy with Polarized Light*(VCH, New York 1986)
199. E. Collett: *Polarized Light, Fundamentals and Applications*(Dekker, New York 1993)
200. R.M.A Azzam and N.M. Bashara: *Ellipsometry and Polarized Light*(Elsevier, Amsterdam 1992)
201. D.S.Kliger, J.W.Lewis and C.E.Randall: *Polarized Light in Optics and Spectroscopy*(Academic Press, Boston 1990)
202. F.A. Jenkins, H.E. White: *Fundamentals of Optics*, 4th edn. (McGraw-Hill, Singapore 1981)
203. B.E.A. Saleh, M.C. Teich: *Fundamentals of Photonics*(Wiley, New York 1991)
204. I.H. Malitson: J.Opt.Soc.Am. **52**, 1377 (1962)
205. T. Radhakrishnan: Proc. Indian Acad. Sci.**A33**, 22 (1951)
206. H. Bach, N.Neuroth: *The Properties of Optical Glass* (Springer, Berlin 1998)
207. J.M. Bennett: Polarization. In: *Handbook of Optics I* ed. by M. Bass (McGraw-Hill, New York 1995)
208. Burr-Brown Corporation: *Application Bulletin (SBOA061): Designing Photodiode Amplifier Circuits with OPA128* (Burr-Brown Corp., Tuscon 1994)
209. J. Wilson and J.F.B. Hawkes: *Optoelectronics, An Introduction* 2nd edn.(Prentice Hall, Cambridge 1989)
210. T.O. Poehler: Detectors. In: *Physical Optics and Light Measurements* ed. by D.Malacara (Academic Press, Boston 1988)
211. S.C. Stotlar: Visible Detectors. In: *Electro-Optics Handbook* ed. by R.W. Waynant, M.N. Ediger (McGraw-Hill, New York 2000)
212. O. Aumala, H. Ihalainen, H. Jokinen, J. Kortelainen: *Mittaussignaalien käsittely* (Pressus Oy, Tampere 1995)
213. G.L. Long, J.D. Winefordner: Anal.Chem. **55**, 712A (1983)
214. P. Horowitz and W. Hill: *The Art of Electronics* 2nd edn.(Cambridge University Press, New York, 1996)
215. G.J. Buist: Stray Light. In: *Standards in Absorption Spectrometry*, ed. by C. Burgess and A. Knowles (Chapman and Hall, London 1981)
216. W. Slavin : Anal. Chem. **35**,561 (1963)
217. A.J. Peyton, V.Walsh: *Analog Electronics with Op Amps* (Cambridge University Press, Cambridge 1993)
218. J. Kauppinen, J. Partanen: *Fourier Transforms in Spectroscopy* (Wiley, Berlin (2001)
219. A. Savitzky, M.J.E. Golay: Anal. Chem. **36**, 1627 (1964)
220. C.Burgess and A.Knowles (Eds.): *Standards in Absorption Spectrometry*(Chapman and Hall, London 1981)
221. H. Mark: *Principles and Practice of Spectroscopic Calibration* (Wiley, New York 1991)
222. E. Vinter: Liquids absorbance standards. In: *Standards in Absorption Spectrometry*, ed. by C. Burgess and A. Knowles (Chapman and Hall, London 1981)
223. D. Irish: Solid absorbance standards. In: *Standards in Absorption Spectrometry*, ed. by C. Burgess and A. Knowles (Chapman and Hall, London 1981)
224. H.E. Bennett: Appl.Opt.**5**, 1265 (1966)
225. A.J. Everett: General considerations on UV-visible spectrscopy. In: *Standards in Absorption Spectrometry*, ed. by C. Burgess and A. Knowles (Chapman and Hall, London 1981)

226. C.J. Sansonetti, M.L.Salit, J.Reader: Appl.Opt. **35**, 74 (1996)
227. N. Fernandes (ed.): *The Book of Photon Tools* (Oriel)
228. J.G. Vinter: Wavelength calibration. In: *Standards in Absorption Spectrometry*, ed. by C. Burgess and A. Knowles (Chapman and Hall, London 1981)
229. L. Ward: *The optical constants of bulk materials and films* 2nd edn. (IOP Publishing, Bristol 1998)
230. K.-E. Peiponen, A.J. Jääskeläinen, E.M. Vartiainen, J. Räty, U. Tapper, O. Richard, E.I. Kauppinen, K. Lumme: Appl. Opt. **40**, 5482 (2001)
231. H. Soetedjo and J. Räty: *Reflectance Study of Contaminant Layer on a Probe Window*, Appl. Spectrosc. **57**, 8 (2003).

Index

Springer Series in

OPTICAL SCIENCES

New editions of volumes prior to volume 70

1 **Solid-State Laser Engineering**
By W. Koechner, 5th revised and updated ed. 1999, 472 figs., 55 tabs., XII, 746 pages

14 **Laser Crystals**
Their Physics and Properties
By A. A. Kaminskii, 2nd ed. 1990, 89 figs., 56 tabs., XVI, 456 pages

15 **X-Ray Spectroscopy**
An Introduction
By B. K. Agarwal, 2nd ed. 1991, 239 figs., XV, 419 pages

36 **Transmission Electron Microscopy**
Physics of Image Formation and Microanalysis
By L. Reimer, 4th ed. 1997, 273 figs. XVI, 584 pages

45 **Scanning Electron Microscopy**
Physics of Image Formation and Microanalysis
By L. Reimer, 2nd completely revised and updated ed. 1998,
260 figs., XIV, 527 pages

Published titles since volume 70

70 **Electron Holography**
By A. Tonomura, 2nd, enlarged ed. 1999, 127 figs., XII, 162 pages

71 **Energy-Filtering Transmission Electron Microscopy**
By L. Reimer (Ed.), 1995, 199 figs., XIV, 424 pages

72 **Nonlinear Optical Effects and Materials**
By P. Günter (Ed.), 2000, 174 figs., 43 tabs., XIV, 540 pages

73 **Evanescent Waves**
From Newtonian Optics to Atomic Optics
By F. de Fornel, 2001, 277 figs., XVIII, 268 pages

74 **International Trends in Optics and Photonics**
ICO IV
By T. Asakura (Ed.), 1999, 190 figs., 14 tabs., XX, 426 pages

75 **Advanced Optical Imaging Theory**
By M. Gu, 2000, 93 figs., XII, 214 pages

76 **Holographic Data Storage**
By H.J. Coufal, D. Psaltis, G.T. Sincerbox (Eds.), 2000
228 figs., 64 in color, 12 tabs., XXVI, 486 pages

77 **Solid-State Lasers for Materials Processing**
Fundamental Relations and Technical Realizations
By R. Iffländer, 2001, 230 figs., 73 tabs., XVIII, 350 pages

78 **Holography**
The First 50 Years
By J.-M. Fournier (Ed.), 2001, 266 figs., XII, 460 pages

79 **Mathematical Methods of Quantum Optics**
By R.R. Puri, 2001, 13 figs., XIV, 285 pages

80 **Optical Properties of Photonic Crystals**
By K. Sakoda, 2001, 95 figs., 28 tabs., XII, 223 pages

81 **Photonic Analog-to-Digital Conversion**
By B.L. Shoop, 2001, 259 figs., 11 tabs., XIV, 330 pages

82 **Spatial Solitons**
By S. Trillo, W.E. Torruellas (Eds), 2001, 194 figs., 7 tabs., XX, 454 pages

83 **Nonimaging Fresnel Lenses**
Design and Performance of Solar Concentrators
By R. Leutz, A. Suzuki, 2001, 139 figs., 44 tabs., XII, 272 pages

84 **Nano-Optics**
By S. Kawata, M. Ohtsu, M. Irie (Eds.), 2002, 258 figs., 2 tabs., XVI, 321 pages

85 **Sensing with Terahertz Radiation**
By D. Mittleman (Ed.), 2003, 207 figs., 14 tabs., XVI, 337 pages

Springer Series in

OPTICAL SCIENCES

86 **Progress in Nano-Electro-Optics I**
Basics and Theory of Near-Field Optics
By M. Ohtsu (Ed.), 2003, 118 figs., XIV, 161 pages

87 **Optical Imaging and Microscopy**
Techniques and Advanced Systems
By P. Török, F.-J. Kao (Eds.), 2003, 260 figs., XVII, 395 pages

88 **Optical Interference Coatings**
By N. Kaiser, H.K. Pulker (Eds.), 2003, 203 figs., 50 tabs., XVI, 504 pages

89 **Progress in Nano-Electro-Optics II**
Novel Devices and Atom Manipulation
By M. Ohtsu (Ed.), 2003, 115 figs., XIII, 188 pages

90/1 **Raman Amplifiers for Telecommunications 1**
Physical Principles
By M.N. Islam (Ed.), 2004, 488 figs., XXVIII, 328 pages

90/2 **Raman Amplifiers for Telecommunications 2**
Sub-Systems and Systems
By M.N. Islam (Ed.), 2004, 278 figs., XXVIII, 420 pages

91 **Optical Super Resolution**
By Z. Zalevsky, D. Mendlovic, 2004, 164 figs., XVIII, 232 pages

92 **UV-Visible Reflection Spectroscopy of Liquids**
By J.A. Räty, K.-E. Peiponen, T. Asakura, 2004, 131 figs., XII, 219 pages

Printing: Mercedes-Druck, Berlin
Binding: Stein+Lehmann, Berlin